A SHEARWATER BOOK

# ACT III *in* PATAGONIA

# ACT III *in* PATAGONIA

## PEOPLE *and* WILDLIFE

---

William Conway

ISLANDPRESS / SHEARWATER BOOKS
*Washington • Covelo • London*

*A Shearwater Book*
*Published by Island Press*



*Library of Congress Cataloging-in-Publication data.*
Conway, William G.
Act III in Patagonia : people and wildlife / William G. Conway.
p. cm.
Includes bibliographical references and index.
ISBN 1-55963-518-5 (hardback : alk. paper)
1. Animals—Patagonia (Argentina and Chile)
2. Human-animal relationships—Patagonia (Argentina and Chile)
3. Patagonia (Argentina and Chile)
I. Title: Act three in Patagonia. II. Title: Act 3 in Patagonia. III. Title.
QL239.C66 2005
333.95'416'09827—dc22 2005002493

*British Cataloguing-in-Publication data available.*

Printed on recycled, acid-free paper
Manufactured in the United States of America

10 9 8 7 6 5 4 3 2 1

*For*

*Robert G. Goelet,*

*who introduced me to Patagonia*

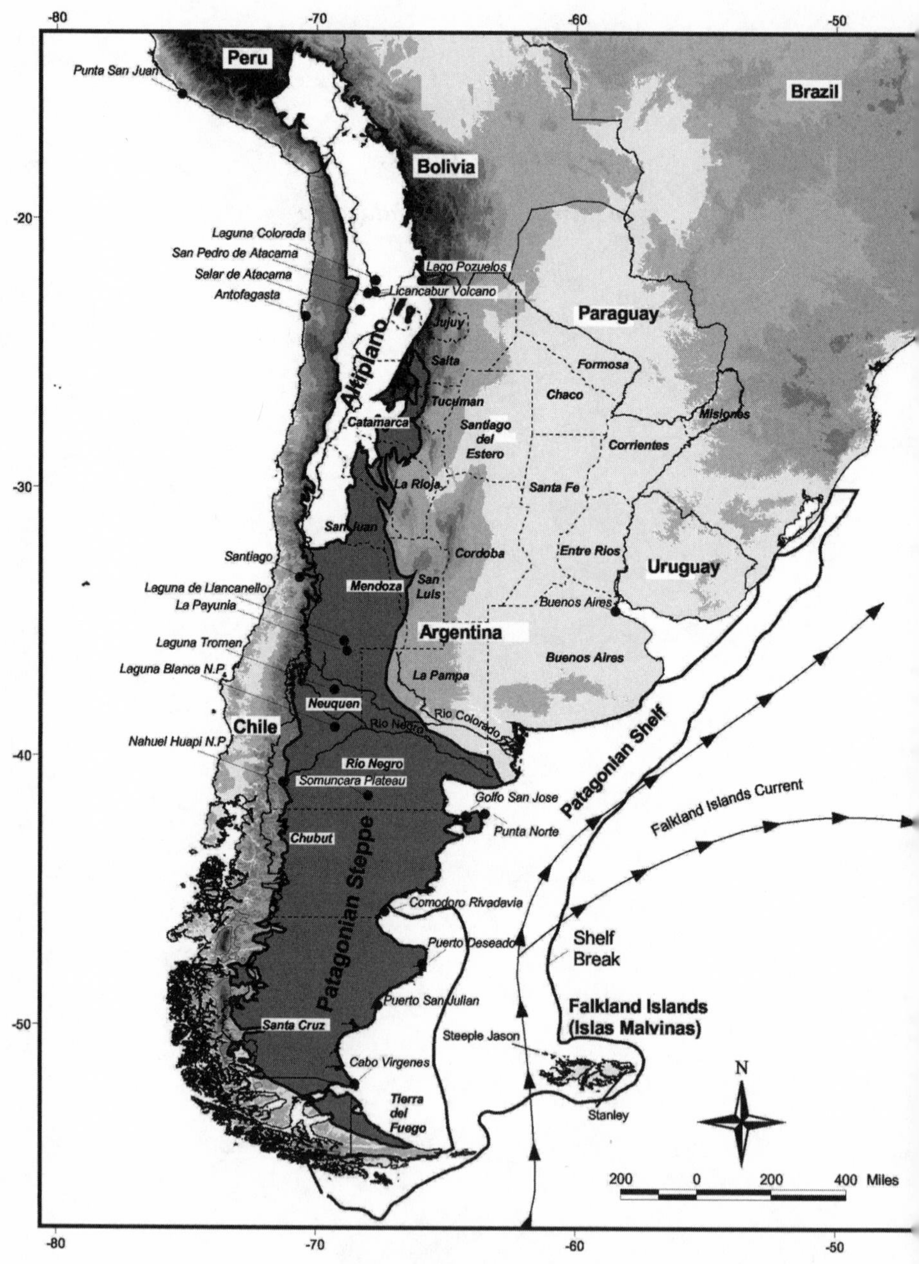


The Southern Cone

# *Contents*

# *Acknowledgments*

Bob Goelet introduced me to Patagonia on three expeditions and has supported Wildlife Conservation Society projects there for more than forty years. His help and encouragement have been extraordinary.

Innumerable treks and conversations, shared discoveries, good food and bad, high winds, grand sights, dusty Christmases, and flat tires have all contributed to this book—most especially, the tolerance and beauty of its absorbing community of wild animals. I wrote with a deep sense of obligation to them.

Graham Harris, a Patagonian, has accompanied my wife, Kix, and me as friend, fellow conservationist, and local expert on countless trips. His intimate knowledge of Argentina provided innumerable insights, and his wife, Patricia, and children, Edward and Sabrina, have become family. The long, close friendships of Felipe and Tessy Lariviere, and of Jimmy Llavallol, have been especially helpful and instructive. As scientists and good friends, Dee Boersma, Francisco Erize, Claudio Campagna, Pablo Yorio, and Andrew Taber played a large part in the stories that follow, and in my understanding of them.

In forty-five years of visits to Patagonia, countless people have helped me, and only a few can be mentioned. Aid in this book's preparation has come from Graham Harris, Dee Boersma, Claudio Campagna, Andres Novaro, Francisco Erize, Pablo Yorio, Felicity Arengo, Ricardo Baldi, Felipe Lariviere, John Behler, Susan Walker, and Liz Lauck. All provided information and read early portions of the text. I appreciate both their wisdom and their patience. I am especially grateful to John Robinson, Graham Harris, Claudio Campagna, and Andrew Taber, who read the entire text and made many important suggestions and corrections. Only a truly outstanding editor could have guided the creation of a book from the wide-ranging material in this one. Jonathan Cobb, of Island Press, has been relentlessly perceptive and unfailingly constructive. WCS landscape ecologist Gosia Bryja kindly prepared the maps.

Over the years, Chubut politician Antonio Torrejon has been a skillful ally, fighting for the future of Patagonian wildlife. Alfredo Lichter, whose educational work is increasingly important to Patagonia, has been a source of inspiration. Helpful in the field in Argentine Patagonia, in addition to some mentioned above, have been Flavio Quintana, Marcela Uhart, Patricia Gandini, Esteban Frere, Alberto and Carol Passera, Roger and Katy Payne, Scott and Anne Swann, and Arturo Tarak; in Peru: Patricia Majluf and Charles and Marianna Munn; in Chile: Michel Durand, Carlos Weber, Mario Parada, Alfonso Glade, Pablo Contreras, Eduardo Rodriquez, and Jürgen Rottmann; in Bolivia: Omar Rocha; in the Falkland/Malvinas: Becky Ingham, Nic Huin, Ben Sullivan, and Rob McGill.

Other Argentines whose help is deeply appreciated include Miguel Reynal, David and Peggy Fenton, Maurice Rumboll, Enrique Bücher, Guillermo Caille, José Luis Esteves, José Maria Musmeci, Alicia Tagliorette, Luis and Doña Mecha LaRegina, Alberto LaRegina, Juan Carlos López, and Alejandro Vila. New York colleagues Frank Larkin, Alejandro Grajal, Billy Karesh, Bob Cook, and John Gwynne have helped and accompanied me on exploratory visits.

Art Ortenberg and Liz Claiborne have been discerning critics and supporters of the WCS Patagonia program since their first trip there, making a major portion of the work possible. Maria Carmen Perez Companc has been both helpful and hospitable, and Amalia Lacroze de Fortabat has long been supportive. The late Roger Tory Peterson and Luis Peña helped to plan and guide, respectively, my first trip to the Bolivian and Chilean altiplano.

Generous support has been provided by the Wildlife Conservation Society and several of its donors, especially Joan Tweedy, Laurance Rockefeller, and the late Nixon Griffis. Fundación Patagonia Natural; Fundación Vida Silvestre Argentina, especially its general director, Javier Corcuero; Centro Nacional Patagónico; Administración de Parques Nacionales; and many officials of the Argentine provinces of Chubut, Rio Negro, Santa Cruz, Neuquén, Jujuy, and Mendoza have been most helpful. The collaboration of Chile's Corporación Nacional de Forestal has been essential, as has that of Falklands Conservation in the Falkland/Malvinas Islands.

Others who have aided the development of the book include John Croxall, James and Elsie Doherty, Steve Johnson, Kent Redford, Victoria Rowntree, Michael and Judy Steinhardt, Doug and Kris Tompkins, and David Western.

The constant help and companionship of Christa (Kix) Conway, my wife, both at home and in the field, made this book possible.

---

Graham Harris kindly gave permission for reproduction of his color photograph of right whales, and the black-and-white reproduction of *Tehuelche Hunting Scene*, painted by John Guille Millais, that appears on the opening page of Act I is from *Through the Heart of Patagonia* by H. Hesketh Prichard (published in 1902 by D. Appleton Company, New York). All other photographs in the book—including that of sea lion skeletons left by sealers on the Peninsula Valdés shore, taken in 1964 and reproduced on the opening page of Act II, and that of Graham Harris and Pablo Yorio at Punta León, Chubut, 1990, reproduced on the opening page of Act III—were taken by me on one or another of my Patagonian trips.

The beauty and genius of a work of art may be reconceived, though its first material expression be destroyed; a vanished harmony may yet again inspire the composer; but when the last individual of a race of living beings breathes no more, another heaven and another earth must pass before such a one can be again.

William Beebe, *The Bird*

# ACT III *in* PATAGONIA

*And when, after the long trip, I arrived in Patagonia*
*I felt I was nowhere. The landscape had a gaunt expression,*
*but I could not deny that it had readable features and that I existed in it.*
*This was a discovery—the look of it. I thought:* Nowhere is a place.
Paul Theroux

---

# Introduction:
## *To Patagonia and the Southern Cone*

In the last pages of *The Voyage of H.M.S. Beagle*, published in 1845, Charles Darwin wrote:

> In calling up images of the past, I find that the plains of Patagonia frequently cross before my eyes; yet these plains are pronounced by all wretched and useless. They can be described only by negative characters; without habitations, without water, without trees, without mountains, they support merely a few dwarf plants. Why then, and the case is not peculiar to myself, have these arid wastes taken so firm a hold on my memory? . . . I can scarcely analyze these feelings; but it must be partly owing to the free scope given to the imagination. The plains of Patagonia are boundless, for they are scarcely passable, and hence unknown; they bear the stamp of having lasted, as they are now, for ages, and there appears no limit to their duration through future time.[1]

On my first visit to Patagonia, to a beach on the coast of southern Argentina that the *Beagle* sailed by in 1833, I was affected by a different experience:

It is a cool, foggy morning. Far out in the Atlantic, hidden in the mist, a penguin brays as two giant petrels fight over one of his kin's carcass rolling in the surf nearby and a huge black-browed albatross sails serenely overhead. Beautifully curled green waves crest and crash on the shore and sharp winds slice plumes of spray from their crowns as the surf seethes high on the beach before hissing loudly over its brightly polished pebbles and back again. A 5,000-pound elephant seal splashes out of the water, raises his barrel-sized head, and bellows a powerful hammering call—a mallet striking a dock. Outlandishly pink, a "V" of eleven honking flamingos appears and flies over him, vanishing in the fog. The ocean's fragrances are wet and sulphurous; it is filled with life. Beyond the beach wrack, under a thin thorny shrub, there is a nest with five clam-sized eggs so glossy that I can see my reflection in them. They are bright green.

Darwin says almost nothing about seals, penguins, and albatrosses and only a little about flamingos. Perhaps this reflects his primary interests in geology and paleontology, but it may indicate the difficulty of finding Patagonia's widely dispersed concentrations of wildlife in such a huge unsettled region in the 1830s.

A vast, scantily peopled region of rough shores, broad steppes, and high mountains, Patagonia, which belongs wholly to Argentina and Chile, makes up the southernmost part of South America (largely below the fortieth parallel). Here, the continent gradually narrows, becoming the Southern Cone; it ends at the bottom of Tierra del Fuego in the forbidding rocky fist of Cape Horn, aimed toward nearby Antarctica. Argentine Patagonia is by far the larger part. It is bigger than Texas and Arizona combined, over a million square kilometers (386,000 square miles), but with fewer than two million people. Many of those hardy people and most of the little-remarked wildlife live along the stormy coast of the southwest Atlantic, a famously dangerous sea covering the largest continental shelf in the entire Southern Hemisphere, twice the size of Argentina. Its waters feed flocks of penguins and albatrosses, herds of seals, schools of anchovies, aggregations of squid, pods of whales, and, now, hundreds of fishing boats—but Darwin said little of this wildlife.

In contrast, the dry shores and arid steppes support herds of camel-like guanacos, flocks of ostrich-like rheas, groups of nothing-like maras, and, nowadays, millions of sheep on scores of estancias. It is this awesomely "boundless" area that dominated Darwin's memories. What rules mine are the wildlife and the unexpected conjunctions of flamingos and camels, Darwin and Drake, the high Andean plains, and even the deep Atlantic abyss. The region easily covers both visions, and, in its preservation, a great deal is now at stake.

Today, many of the Southern Cone's strange assemblages of wild animals live in frontier-like associations with settlers, who both use them and are unsettled by them. Most of these people are economic immigrants with no history on the land or with its wild creatures. They don't know the animals. In some places, the human population has quadrupled in the last thirty years, and the larger species have been decimated. Economic globalization is imposing a pernicious kind of exploitation, in which non-local investors are influencing land use, development, and wildlife harvesting. Scientific studies and wildlife tourism, however, are beginning to create strong counterforces on behalf of conservation. Environmental awareness is growing, and, most encouragingly, most powerfully, it is inspiring new collaborative plans, supported both locally and internationally, that promise long-term preservation solutions for the coast and the sea, and even restoration for the boundless plains.

My armchair interest in the Southern Cone began with Darwin's writings when I was a teenager but became real with an unexpected invitation in 1960. I had just returned from an expedition to the high (14,000+ feet) salty lakes in the Chilean and Bolivian Andes, at the northernmost edge of the Southern Cone, when I was invited to visit Argentine Patagonia. The grand vistas, unforgiving environments, and scarcely mentioned wildlife that I saw captured my imagination. From the standpoint of wildlife survival, parallels with the oxygen-poor Andean altiplano (high plains), were unmistakable, yet refreshingly unique and bizarre. At the same time that pairs of Patagonian penguins courted in waddling dances by a fierce sea, colonies of altiplano flamingos were strutting shoulder to shoulder in inhospitable

Andean lakes over two and a half miles higher. I also came to realize that, whether rearing their young on Patagonian steppes, Andean altiplano lakes, high cliffs, or distant oceanic archipelagos, the most fascinating of these Southern Cone creatures are those that breed in large groups, or colonies, and are dependent for their survival on an island-like inaccessibility.

A curator of birds at the Bronx Zoo when I first visited Patagonia, I became director of the zoo in 1961 and, eventually, president of its parent, the Wildlife Conservation Society (WCS). The society now operates four New York City zoos, an aquarium, and an international conservation program, which became a major focus of my efforts. The collaboration of a committed cadre of board members and staff soon resulted in the initiation of more than 350 wildlife conservation projects in fifty-two countries. About 18 of those projects are in the Southern Cone, where I take a special interest. Almost all of those 18 are run by first-rank local scientists who were identified on my frequent forays to Patagonia, which I superimposed onto my New York duties. Most have become staff members or grantees of WCS. They have produced hundreds of contributions to conservation science and stimulated far more news stories and television documentaries—and, most important, new wildlife reserves, conservation practices, and laws. In 1999, I stepped down as WCS's CEO, but I continue my work in the Southern Cone as WCS's "senior conservationist." A rough calculation reveals that I have driven over 150,000 kilometers in the often rugged terrain of Bolivia, Peru, Chile, and, especially, Argentina, seeking opportunities to promote wildlife and learning the land firsthand—and logging forty-nine flat tires, so far.

All the Southern Cone places I know are alike in having seriously wild weather, though each is different in its mixture of cutting winds and cold, heat, storms, and drought, and human-wildlife associations. In this book, their geography includes Argentine Patagonia, except its southern mountains and forests (where wild animal numbers and diversity are comparatively low); the largely uninhabited puna and altiplano of Andean Bolivia, Argentina, and Chile; the long Atlantic coast of Patagonia; and the waters of the south Atlantic out to the Falkland/Malvinas Islands.* These places are neither ordinary nor superficially similar, yet they all support, and are connected by, a suite of sizable birds and mammals that have been heavily exploited by people. Most of these creatures are colonial, and all have outlasted the glory days of Southern Cone fauna, twelve thousand years ago, when the land was inhabited by sloths as big as rhinos, giant camels, mastodons, and strange horses. They have also outlived the ancient Foot Indians from the same period. Neither those peoples, nor the more recent Tehuelches, who were discovered by Magellan and interviewed by Darwin, raised crops, domesticated animals, built permanent structures—or survived Western settlement. They never really ventured into the Atlantic, so they never found the Falkland/Malvinas, but they changed wildlife—permanently. I write about them in Act I of this book.

Since the first days humans ventured onto the steppes and down to the south Atlantic, humanity's role in the Southern Cone has occurred in a sequence of three major acts, as I conceive of it. First the mega-creatures, the giant sloths and glyptodonts, vanished, as did the wild horses and the Foot Indians. Act I was over. Next came the extermination of the Tehuelches, in the nineteenth and twentieth centuries, along with widespread wildlife butchery. Coastal mammals, whales, and steppe animals alike were slaughtered, and much of the steppe's fragile vegetation was destroyed with the introduction of sheep and hares. We could call that annihilative phase Act II. The slaughter of sea mammals is mostly behind us now, and overgrazing is in uneasy stasis. Except for extensive overhunting and overfishing, the combination of ignorance and abuse that characterized Act II is sputtering toward an ignominious end. Much of the Southern Cone is now settled, but there is still lots of open land. Conservation science and public interest in the environment have grown, and imaginative new conservation initiatives are under way. Act III has begun, and there remains a big, strange, natural world worth saving if it can be more generally

* Falkland/Malvinas: Both Great Britain and Argentina claim this archipelago.

understood. This book is an attempt to increase understanding. It is about this region's wild creatures and their relationship with people, and how the land's lonely beauty, the wildlife's singularity, and the region's natural diversity might be saved.

---

The circumstance of a developing modern society living cheek by jowl with penguins, camels, flamingos, whales, and the world's largest seals in the once remote Southern Cone inevitably leads to conflicts and interdependencies that are often dramatic and sometimes lunatic. The situation is fascinating enough and the outcome unclear enough that it keeps bringing me back to promote on-the-ground field science and conservation. Field science is the process of figuring out how nature works, and conservation is figuring out how people can fit into the works without wrecking them. Each year, studies of the Southern Cone's wild animals divulge more about them and about the people who live with them. Although the details of such accounts are local, their implications are global. They not only reflect the newly complicated situations of wildlife over much of the world but also offer fresh insights.

Simple stories are easy to communicate. No matter how wondrous and elegant, complicated stories are not. Almost everywhere, wildlife protection is handicapped by the incomprehensible complexity of most natural ecosystems. While astronomers can predict eclipses of the sun with a sureness measured in fractions of a second, conservation biologists cannot accurately predict the extinction of an animal species that lives in the woods next door. However, in contrast to the astrophysicist and his eclipses, the conservationist can sometimes do something about potential extinctions.

The conservation stories of the Patagonian coast and steppe, the Andean altiplano, and the Falkland/Malvinas are not the sadly usual ones of lumbered trees and vanishing species. There are almost no trees. Nor are these stories about the hopeless loss of species by the thousands—jewel-like beetles and butterflies, monkeys, pythons, and rhinos. There are relatively few species. The Southern Cone's remaining biodiversity is not so much in tree-filled forests as it is in pristine shores and arid bushland, rocky islands, salty lakes, and nearly plantless plains; and that is where the main conservation stories are taking place. It is in eye-filling animal cities of sea lions and seals and penguins and terns and gulls and cormorants breeding amid the dusts of a desert shore but living on a cold ocean's bounty as much as its fishes do. The singularity of the Southern Cone's diversity is also in the curious grace and loneliness of sentinel-like guanacos gazing into the vast steppe from fossil-laden clifftops and in little families of dinosaur-like rheas grazing on leathery quilimbay and coarse stipa on the windy plains below. And it is in high cliff towns of burrowing parrots, in exquisitely specialized colonial flamingos dwelling among Andean volcanos, and in long-winged albatrosses gliding far out to sea and nesting in the remote Falkland/Malvinas Archipelago.

Unraveling the natural histories of a region's cast of animals is the first step in understanding the drama of its conservation. It is crucial to the understandings that must support the conservation efforts to come. Years must be spent wresting crucial facts from suspicious wild creatures. Who would have guessed that sea lion bachelors join forces to kidnap females from "beachmaster" harems? Or that a Patagonian mouse as big as a Scotch terrier breeds in communes but mates for life? Or that the males of certain ostrich-like birds incubate the eggs and rear the chicks, that there are whales with testes bigger than couches, and that the camel-like guanacos hum while mating? As a result of our studies, for example, we learned that forty-one thousand penguins were dying each year from oil pollution, that black-browed albatrosses were being killed by commercial fisheries at the rate of several each hour, and that every egg of the world's rarest flamingos was being eaten by villagers. Such facts are the foundations of conservation.

In Patagonia, wildlife conservation is developing in a society of new immigrants, people midway between frontier gaucho and TV-cultivated cowboy. But conservation puts society's interests above those of the individual entrepreneur, or even the individual community. That makes it necessary for people not only to value wildlife but also to become well informed about what its preservation entails, to think about how wildlife relates to the land and the sea, to habitats and ecosystems, and to other wildlife—not just about individual animals and species. The fortunes of the Southern Cone's wildlife are likely to be played out in those of a suite of creatures whose needs define the habitat necessities for almost all the others, the big "landscape species." These species' survival requires the largest areas, and their conservation, thereby, has the greatest positive impact on biodiversity as a whole. Do what it takes to preserve toothfish, penguins, guanacos, and flamingos, and you will save most of the rest. Worldwide, more than one hundred thousand protected areas covering 19.7 million kilometers (7,604,200 square miles) have been set aside—on paper—to preserve some portion of nature for "future time." But that is not nearly enough; too many are in the wrong places, and most are not protected except on paper. Nevertheless, of all the many things people do to save nature, establishing big protected areas is the most powerful, the most lasting. They are not all that's necessary; they're simply the most effective. But anyone who thinks that saving wildlife is therefore straightforward is wrong. Human ignorance remains a large part of conservation's problem. We too willingly accept quick and easy wrong answers and romantic ideologies over longer, harder, right ones. Examples of scientific and environmental illiteracy can be wonderfully astonishing—here are two of my favorites:

The late Sir Peter Scott, deeply missed dean of conservationists, claimed to know an English paleontologist whose least becoming suspicions were confirmed by an experience in his public rose garden in the Cotswolds. The garden became a favored tourist stop when the paleontologist paved its paths with casts of dinosaur footprints: a *Stegosaurus* track before the hybrid teas, *Triceratops* near the floribundas, *Tyrannosaurus* impressions before a fine "Paul Scarlet," and so on. Visitors loved it. One evening, however, as the last minibus of enthusiasts prepared to leave, one troubled guest hung back. "The thing that surprises me," she finally confided, "is that they would come so close to the house."

George Gaylord Simpson, the famous paleontologist, enjoyed relating a conversation he had with the wife of a Patagonian estancia administrator. "I told her that we had found a fossil crocodile in the barranca below the house, and she said, 'Oh, sí, sí, a crocodile,' as if she had to shoo them out of her kitchen every morning.

"'Yes,' I said, 'a very ancient one, too. It is over forty million years old'—a great understatement, but I wanted to let her down easily.

"'And is it dead?'"

Dr. Simpson adds, "I know Mark Twain had a similar story long since, but this is perfectly true."

These anecdotes probably tell us as much about the cynicism of scientists as they do about the layman's biological naiveté. Nevertheless, large numbers of people understand neither their own relationship with grass, cows, and chimps, nor what Tim Flannery calls "the eternal rules of ecology which bring order and meaning to our lives."[2] Many stand disabled before such essential concepts as environmental "carrying capacity" and are easily misled not only by others just as much at a loss but also by disinformation put forth by special interests. The need for environmental literacy out of self-interest, let alone wildlife conservation, cannot be overstated.

My experience of the human-wildlife interface in zoos and in the field gives me little confidence in an intrinsic and universal human sympathy for wild animals, a "biophilia."[3] I should be able to claim, therefore, that the scientific survival imperatives of conservation alone impel my own fascination with nature. Instead, it is, I confess, its matchless beauty—and biophilia. It sobers me that others of similar background often feel differently.

Wild animal preservation falls into four main categories: First, saving habitats; second, protective laws, policies, and actions; third, protective human cultures and economies; and fourth, in extreme situations, intensive care and management—exceptional ways of effecting the first three. That understandable options for such preservation exist in Patagonia fuels my hopes for the Southern Cone.

Once the facts are in hand, wildlife preservation is far more a social and political process than a scientific one. It becomes a matter of learned values, information, consensus building, and analytic deliberation, and it struggles incessantly for a foothold on the slippery slopes of ethics and aesthetics. Ultimately, conservationists face the task of preserving the environment by providing facts and trying to inspire a concern for nature so compelling that it will win high priority among human hopes and needs, even without immediate economic reward and even on the distant frontiers of the Southern Cone.

If we do nothing, Act III will end the Patagonian drama as a tragedy. Its closing curtain will fall, little by little, greedy step by thoughtless blunder, to a conclusion of irreparable silence. And wild Patagonia will be nowhere. But good things are being done, new voices are being heard, and I will tell the story of imaginative conservation efforts well under way. Act III in the relationship of Southern Cone humans and wildlife is different from the past. Its outcome could be a novel stewardship of the land, the waters, and their wild creatures, one that is sustainable. It could conclude with an ovation of human support, a wealth of understanding, and with curtain calls for Act III's public officials, scientists, farmers, fishermen, industrialists, and wildlife constituencies. And wild Patagonia will then survive.

*All the attending marvels of a thousand Patagonian sights and sounds helped to sway me to my wish.*
Herman Melville

# Prelude

It was August 7, 1810. A band of tall Tehuelches crept down a dry cañada and through the thorny gray bushes along the shore of Peninsula Valdés' shimmering Golfo San José, only a few miles from the present-day site of a Wildlife Conservation Society research station ("whale camp"). Silently, they surrounded a crude little church.

For more than thirty years, since 1774, a small group of Spanish friars had lived on the shore of the Golfo under extraordinary conditions of loneliness and deprivation. While they were in church that August day, the Indians attacked and massacred all but one man, who somehow managed not only to escape but also to get to Viedma, more than three hundred miles away, and tell the tale. It would be another seventy-two years, long after the Spanish had abandoned their Argentine aspirations, before the first non-Indian would successfully settle on Peninsula Valdés. He was Gumersindo Paz, a tough, broad, black-bearded Basque from Viedma, who brought his family to live, at first, in a corrugated metal lean-to.[1]

The hatchet-like shape of the peninsula on which Paz made his home dominates maps of the Patagonian shore, and at 2,404 square kilometers (1,500 square miles), it is big enough to be noticeable even on world maps. Like much of Patagonia, the high bluffs, broad plains, and rolling bush lands of Valdés were once the floor of an ancient sea. They are strewn still with Miocene flotsam, including oyster shells a foot wide named Hatcher's oysters after a Princeton geologist who worked in and loved Patagonia. The old seabed is at least eleven million years old, and its leftover debris includes amazing masses of sand dollars and ancient mollusks. The plains are covered with coarse but often beautiful grasses such as fletchillas and stipas, and tough scratchy shrubs: zampa, quilimbay, jume, molle, uña de gato, piquillín, and jarilla. Called creosote, jarilla (*Larrea* spp.) made it to the North American west only ten thousand years ago, and some think it came all the way from Argentina, perhaps with migrating birds.[2] Near the sea, the whitish wind-eroded cliffs are ornamented with twisted, bonsai-like shrubs, especially falso tomillo, and the shores are famously covered with beautifully colored pebbles, "Tehuelche gravel."

Patagonia's pebble paving is the most extensive on Earth and one of many unsolved mysteries of the place.[3] The stones measure up to four inches in diameter and are often finely polished—bright red, orange, pink, purple, green, yellow, black, brown, white, and endless combinations of those colors. They are not from anyplace nearby. Some experts believe that the pebbles are the result of thousands of years of tumbling in streams down from the distant Andes, receiving their final polishing in the surf of ancient shores. They are mostly made of porphyry, granite, petrified wood, quartz, and basalt.

## Counting Elephant Seals on Peninsula Valdés

In 1964, when the peninsula was little visited and the roads were sometimes impassable, Bob (Robert G.) Goelet, a WCS trustee and longtime Argentina enthusiast, brought Bill Drury, an avian ecologist from the Massachusetts Audubon Society; my wife, Kix; and me to Peninsula Valdés. Our immediate objective was to finish *From the Pampas to Patagonia*, a wildlife film that Bob and I had begun in the pampas a few years before to create interest in Argentina's wild animals. It became one of the first nature films made in Argentina, where it was widely shown.

Bob was tall, bespectacled, bone thin, and a former navy pilot; his qualifications for travel in Argentina included an ability to drive either truck or car a dozen hours at a time without relief—often with aeronautical velocity. Despite our dirt-road filthiness and meager language skills, we were welcomed everywhere. (Bob had an alarming collection of Spanish words, fluent French, and undaunted verbal courage. Dark-haired and mustachioed, I was commonly taken for an Argentine but had the least Spanish. Kix, like many Europeans, speaks in tongues, however, two of her best being Spanish and German.)

A type of huge seal had attracted Bob to the shore—the southern elephant seal, named for its great size and the males' large rubbery noses. He had read that some elephant seals could still be found on Peninsula Valdés, at Punta Norte, in an area where vast rookeries of sea lions once bred. If we could find them, they would be a highlight for our movie. Unfortunately, Punta Norte was nearly a thousand miles south of Buenos Aires on the mostly dirt roads of the time. A line drawn due east from Punta Norte, across the Atlantic, would pass two hundred miles south of Africa's Cape Town. So the peninsula is about as far south of the equator as New York City is north, and, because of the narrowness of its attachment to the mainland, it is nearly an island.

Elephant seals are the strangest, most remarkable creatures I know—despite my zoo experience. In the first place, they are huge, the largest of all living seals, and they have almost supernatural diving ability. There are two species: the northern elephant seal (*Mirounga angustirostris*—named for its especially big nose), which lives along the coast of California and northern Mexico, and the southern elephant seal (*M. leonina*), which lives mainly in Antarctic waters and on subantarctic islands. The Peninsula Valdés elephant seals constitute the only continental colony of southern elephant seals in the world, and they are even larger (except for their noses) than the northern species. An adult male can measure 21 or even 22 feet long (18 feet, or 5.5 meters, is average) and weigh, it is claimed, as much as eight thousand pounds, though six to seven thousand pounds may be closer to the truth. In any case, the species is, after the elephant itself, the largest of all mammals to produce its young on land, twice as large as the walrus.

Bob's eye had been caught by a report from a distinguished Argentine biologist, Santiago Carrara, who had found a population of elephant seals at Punta Norte that he estimated to number 115 in 1952 and 220 when he returned in 1954. Now, ten years later, we wanted to count once more. All other continental colonies had been wiped out, and the remaining populations lived on remote islands. The seal's blubber yields high-quality oil, which was used in making soaps and paints and, in earlier times, in house and street lighting. Between 1809 and 1829, for example, the population on Macquarie Island (near Australia) was reduced from about 100,000 to 33,000, and on many islands, populations were completely exterminated.[4]

Thanks to an old friend of Bob's, we were invited by the late Emilio Ferro to stay at his handsome Valdés estancia, La Adela, within easy reach of Punta Norte and a decided improvement on Paz's tin lean-to. When we finally arrived at the Punta, we were appalled.[5] The beauty of the vast beach and uninterrupted horizon was eclipsed by acres of bones and mummified carcasses. It called to mind a battlefield where the fallen had been stripped of their uniforms and their bodies left to rot—uncovered graves on the shores of Gallipoli. There were thousands of skeletons and piles of bone fragments. Remaining bits of skin and flesh were black and brittle, the skeletons dry and bleached. A desiccated pile of sea lion skins still stood at the top of the beach, stacked neatly next to a series of slime-filled pits where sea lion or elephant seal oil had apparently been stored after rendering. One could not help feeling disgusted and embarrassed as well as saddened—so much slaughter for such trivial purposes. Almost all of the remains I could identify were of sea lions, and their state of deterioration confirmed, as we had heard, that the killing had ended several years earlier.

Happily, we soon found live elephant seals—you couldn't miss them. Not only are they huge, but they sleep on the beach in groups and are as audible as they are visible. Only a few adult males were present. The procreative season was over, and the harem-masters had gone to sea to feed and recuperate from their long fast during breeding. Many adult females were there, however, plump and matronly, along with young males, mostly molting into new coats. Their old fur is shed in pieces and leaves them looking like fat rag bags. There were also "weaners," this year's babies, weaned by their mothers and left on their own to whatever prospects their imminent entry into the dangerous cradle of the sea would bring.

The faces of these young are unforgettable—large, brown, saucer-like eyes and gentle expressions. Their muzzles have vibrissae, whiskers, like a cat. I counted thirty-one on each side on one youngster and five above each eye, a sensory array that must be very useful in turbid waters. The big eyes are set in a small oval face, which can look rather reproachful, and the relatively small head merges with a 300-pound (136-kilogram) balloon-like body of soft ivory gray. A Wodehouse sentence puts it perfectly: "[He] . . . looked as if he had been poured into his clothes and had forgotten to say, 'When!'" A more appealing creature is hard to imagine, the more so as one realizes that, like most young wild animals, the great majority will not long survive. But the presence of many youngsters was encouraging, for it indicated a healthy breeding population.

To update Carrara's count, Bob and Bill headed west along the pebble beach, and Kix and I headed east. From time to time, an immense male would rise up and, with head five or even six feet in the air, inflate his bulbous nose and emit his resonant thumping call. Surprisingly, it is wild and thrilling, powerfully percussive, appropriate to a pile driver, and the only really respectable sound the species makes. When disturbed by another sea elephant—or a biologist—the usual response is a belching, hissing roar or a growling sort of gurgle.

Sleeping elephant seals are a noisy lot, too. They breathe stertorously but normally enough for a while and then stop breathing altogether for ten minutes or more, a sleep apnea that must serve them well at sea. If an observer gets very close, he may regret it, for the *elefante* may resume breathing (about the time you fear he has died) with a gooey snort. At any time, elephant seals are likely to emit a great many sounds that fall coarsely on human ears. Much is explained by the studies of R. M. Laws, who measured the intestine of a bull elephant seal and found it to be 662 feet long, forty-two times that particular animal's body length.

Although we had missed the breeding season, we had not missed all of its contests. I was watching a group of pigeon-like snowy sheathbills

(*Chionis alba*) nonchalantly strolling between the sleeping monsters and picking at bits of shed seal skin, like track cleaners between locomotives in a switching yard, when suddenly, for no apparent reason, a bull snorted and reared up with a roar. A bigger bull, perhaps five feet away, reared and roared in response. A slow advance began, heads turning, mouths open wide, a staccato call—then a feint by one, and a response by the other. Suddenly, the larger bull lunged, delivering a downward blow with his short tusks executed by the whole front part of his body. The smaller bull struck back, and, as I automatically counted while film clicked through my Bolex, four seconds of rapid, blubber-shaking blows were thrown. The smaller bull was pushed back until he raised his hindquarters, lowered his head, and, with mouth open defiantly and many a backward glance, retreated down the beach. The larger bull slowly sank onto the warm shingle and went back to sleep.

With care, keeping one's profile low, it is sometimes possible to approach very near a sleeping seal to get a close-up photo of the face or flipper. Bob had the temerity to tickle the posterior of one sleeping youngster, and the response was delightful. First, a lazy, remarkably handlike flipper reached back and scratched slowly, at least two feet short of the spot. Bob tickled again. Junior itched again. Suddenly he awoke and looked back with the most wonderful expression of outrage and consternation. Assessing the situation, he swung around and faced Bob with a hiss and widely opened mouth—and we slunk guiltily away.

Scratching is a major occupation for elephant seals. Their short, handlike front flippers are very different from the relatively inflexible elongated flippers of sea lions. Evolution has minimized the elephant seal's back legs and positioned them so far rearward as to be of relatively little use for anything but swimming. Its front limbs, more arms than flippers, are unexpectedly prehensile. An elephant seal can rub his chin, scratch his back, clean his nose, pick his teeth, and even—I've photographed it—give the appearance of stifling a yawn. There is an engaging improbability in the combination of a whale-like animal with such slender and dexterous "hands and arms."

Our walk down the eastern sector of the beach had extended to about three and a half miles, and it was time to turn back. Although our count had not exhausted the shore of elephant seals, and we had not attempted to estimate the occasional seals that appeared in the water, the major aggregation seemed behind us. My notebook tallied 1,201, a number more than five times that of Santiago Carrara's 1954 estimate, even without Bill and Bob's figures from the western sector of the beach. When we put our figures together, we came up with a total of 1,387 actually counted and a population estimate of at least 2,000. Punta Norte's herd seemed on its way to becoming a vigorous colony.

Today, forty years on from that walk on the beach, most of the bones have long since been washed away, and now thousands of tourists visit Peninsula Valdés every year. The elephant seals are carefully protected on their beaches by the province of Chubut, and the Valdés animals—numbering now more than forty-five thousand—are thought to be the only elephant seal population still increasing anywhere in the world. Our visits, articles, and movies—and, especially, the hard work of the Argentine officials they helped to convince—played, I believe, a major part in the resurgence.

No better introduction to the wildlife of Patagonia, its fascinations, problems, and possibilities could have been arranged than counting elephant seals on Peninsula Valdés. And no better creature. Amazing things have subsequently been learned about elephant seals. Yet they remain mysterious beasts with powers that still surpass our understanding—like much else in Patagonia and the Southern Cone.

I was hooked by that first encounter. There were thought-provoking parallels with my experiences in the austere altiplano of Chile and Bolivia, and I soon began spending some part of almost every year learning the landscape, the wildlife, and the human connections. The Southern Cone, especially Patagonia, is not only a place where new things can be learned and a wide array of wild animals protected, but also a place where patterns can be discerned that can inform our understanding of how wildlife and humans live together in other parts of the world. The Southern Cone's habitats are populated by creatures that have found ways to flourish on edges and margins, where more and more wildlife is being forced to live everywhere nowadays. In this sense, their unusual stories are universal.

In 1964, the vast majority of the remarkable wildlife from the high

Andean lagunas to the southernmost shores and seas of the southwest Atlantic was neither appreciated nor protected. Most Patagonians were uninterested, there was no ecotourism, and only a few scientists were seriously involved. Each generation of wild creatures was subject to the short-term whims of the latest ungoverned entrepreneurs. Conservation was rare. Waiting in line with other victims of Patagonia's bad roads at a *gomeria*, a tire repair shop, on an early trip, I found myself discussing wildlife with a local estancia manager, who observed: "You could put all the conservationists in Patagonia in your old Ford and still have room for a large dog in the back seat."

## Subplots and Stage Directions in Argentina

Conservation was a decided afterthought, if thought about at all, in the politics of post–World War II Argentina. The 1960s, on through the 1970s and into the 1980s, were thus certainly not an easy time to conduct wildlife studies and work on conservation there. With exquisite timing, an economically strong Argentina had declared war on Germany seven days before the latter's defeat in World War II. However, the period immediately after the war saw the nation's financial advantage dwindle rapidly, squandered by the populist government of the time, which cast a long shadow over the ensuing years of Argentine history.

The first two decades following my first visit to Peninsula Valdés comprised the interregnum between Juan Perón's ouster as president in 1955 and his short-lived resuscitation in 1973; the government of his conspicuously unqualified widow, "Isabelita" Martinez de Perón; and the military dictatorship that ousted her and remained in power until 1983. The period included financial disasters, kidnappings of many local people and foreign corporate officials, the massive and bloody internal assassinations and "disappeareds" of the so-called dirty war (1976–1983), and, in 1982, the tragic campaign to capture the Falkland/Malvinas Islands from Great Britain. Then, as now, Argentina operated in the shadow of the curious cult of Juan Perón and Evita, his revered second wife. The largest Argentine political party remains the often disruptive Peronista; no other national party has been able to complete its term in office in more than forty years.

After the Falklands war, when defeat and reason finally returned the military to its barracks and a democratically elected president came into office, troubles of a different sort arose: Inflation quickly climbed to 4,000 percent a year, for example. My Argentine friend Jimmy Llavallol wryly observed, "A new shirt costs more pesos today [120,000 in 1982] than my home near Bariloche cost in 1930." Kix and I breakfasted at a fancy Buenos Aires hotel for 390,000 pesos ($6.29) and acquired a Ford pickup truck for penguin research in Patagonia for the sum of 550,000,000 pesos ($8,800). To obtain all those pesos, a friend took Kix to a bank (using the back door rather than the front to avoid the ridiculously overstated official rate), and she brought the money back to our hotel in a bulging shopping bag. Eight years before, the exchange had been 8 pesos to the dollar; now it was 62,000. It was a terrible time for Argentines and, consequently, for Argentine wildlife. Wild animals were meat, sport, pests, or inconsequential.

During the 1980s, every half-baked scheme any entrepreneur could think of that might winkle a peso out of leaderless government agencies, so often the last resort of the politically astute, or put food on the table, no matter the costs to one's ethics or the nation's long-term viability, was likely to be attempted. That attitude continues to be an obstacle to rational management in much of the country today. By accident, it has also meant that foreign scientists who might bring in some income or positive attention are usually welcomed—though Argentines are wonderfully welcoming under almost any conditions.

Finally, in the 1990s, by equating the peso with the U.S. dollar and privatizing many badly run government enterprises, President Carlos Saul Menem brought financial stability to Argentina. The country's inherent wealth overcame its generally sanctioned mismanagement—for a while. Those promising days passed all too soon, and by the new millennium, bad times had returned; Argentina was again in serious financial trouble but again appearing to recover—though with many more impoverished people than previously and still the enormous difference between the rich and everyone else, a characteristic shared with most Latin American nations. For a country so generously endowed with agricultural land, livestock, and natural resources, and a people with a notably higher literacy rate than that in most other Latin American countries, its condition is frustrating as well as sad.

The Republica Argentina is South America's second largest nation in both space (1,055,400 square miles; the U.S. has 3,535,000 square miles) and people (thirty-eight million, twice the population of Australia and 80 percent urban). It is not only a land of great natural beauty and diversity of landscape but also one of the world's five major wheat producers.[6] Its famous cattle herd is fifty-five million strong, and the country is nearly independent in petroleum. With fifteen hundred wineries, it is the fourth largest producer of wine in the world. And in Latin America, it is a major tourist destination. Las Leñas, near Malargue, is the largest ski area served by lifts in the Western Hemisphere—bigger than North America's Whistler/Blacomb, Vail, and Snowbird combined. Yet once again, in December 2001 and January 2002, riots rocked the country as the government's continuing bad management led to economic crisis. Five presidents took office in twelve days. The value of the dollar-pegged peso declined 50 percent in a few weeks and to only 28 percent of the dollar in a few months. In November 2002, Argentina defaulted on its IMF loans, the largest such default ever.

Néstor Kirchner, governor of the scantily populated but magnificent Patagonian province of Santa Cruz, was elected president in March 2003—to the surprise of many. He immediately began a program of reform, though whether accountability will truly be introduced into Argentina's government remains to be seen. By early 2004, unemployment was still over 16 percent. Some public services, including national parks, were largely dysfunctional, and, especially in Buenos Aires, kidnappings and robberies—with the police themselves sometimes implicated—had greatly increased. Unfortunately, natural history and conservation cannot ignore unnatural history and lack of effective government.

The reputation of Argentina's roller-coaster economy, ill-considered war with Britain, and, it must be admitted, corrupt officials has submerged the attractions of spectacular Buenos Aires, now a megalopolis of more than twelve million (up from five million in 1950), with its Teatro Colon (twice the size of Milan's La Scala), the Calle Florida, and even its bucolic pampas, not to mention its spectacular coast, high Andes, and lonely steppes. In world affairs, Argentina has been pushed off the map.

And Argentina's Spanish founders were pushed off the South American continent two centuries earlier, swept away by a cadre of Latin American independence movements and subsequent immigration. In 1808, Spain controlled an area stretching from California to Cape Horn. Seventeen years later, she retained only Puerto Rico and Cuba.[7] In 1810, when Tehuelches massacred the Peninsula Valdés friars, there were 500,000 Argentines; by 1914, there were at least 7,885,000. Already, the population of the new people was probably an order of magnitude greater than that ever achieved by the Tehuelches and related tribes. Immigrants poured in and settled, overwhelmingly, in the cities. Sixty percent of the population today is of Italian heritage; the next largest percent is of Spanish heritage. There are also scattered enclaves of Germans and English. The most significant indigenous populations today, many of their number having been deliberately killed, live in the Andean and Chaco provinces; although growing numbers of landless Bolivians, Chileans, and Brazilians are building a sizable underclass throughout Argentina.

In Patagonia itself, during much of the twentieth century and still today, the land remains sparsely populated, its economy tied to sheep ranching on the steppes, fruit growing in the Andean foothills, and oil (after its discovery in 1907), although fishing and manufacturing are growing. By the 1970s and early 1980s, when WCS-supported studies of whales, penguins, and sea lions got under way, Patagonia had begun to change in a host of ways. Industrial developments little related to the desert or coastal resources, except for ports, appeared and attracted more immigrants from Argentina and abroad. Within twenty years, coastal cities doubled in size, then trebled and quintupled, and wildlife and nature found the threats of development overtaking those of hunting. Nevertheless, Patagonia's population is still under two million, and 75 percent of the people in the region live in just thirteen towns and cities on the coast. In fact, less than 5 percent of the entire nation's population is rural. Over 95 percent of the land is privately owned, which constrains the potential for creating protected wildlife areas.

Although the Tehuelches are long gone or assimilated in Patagonia, they seem to have bequeathed to many of their European successors an underlying distrust of the *Porteños*, the people of Buenos Aires, who so dominate the nation. "Yo soy Patagónico" ("I am Patagonian") can conclude almost any conversation about provincial-federal relationships in Patagonia—especially if its way is lubricated with a few *cervezas* and a willingness to turn a blind eye upon the sources of provincial monies, much of which comes from the federal budget. An almost complete disconnect exists between national and provincial authorities insofar as wildlife protection is concerned. Each of the twenty-four provinces into which the country is divided decides on the use of its own natural resources, and many have seen national parks not as an asset but as a withdrawal of land from local use and revenues. Except for the reserves the provinces themselves have created and one at Monte León in Santa Cruz, there are no national parks or reserves on the entire spectacular Patagonian coast. The national park at Monte León came about in 2001, thanks to U.S. philanthropists Doug and Kris Tompkins (pursuant to planning by former Argentine National Park presidents Felipe Lariviere and Francisco Erize). Moreover, many of the big national parks that have been created in the Patagonian Andes have been the targets of repeated grab, squeeze, and plunder efforts by provincial politicians, mostly unsuccessful thus far. Given Argentina's distinguished national park history, this seems humiliating. After all, Argentina was the third nation in the world to establish a national park, Nauel Huapi, in 1904. The United States' Yellowstone in 1872 and Canada's Banff Park in 1885 preceded it.

So far, the major species of Argentina's wildlife have managed to survive gross overhunting and increasing encroachment on their habitats, but only in ever smaller, more threatened populations. A century after the gazetting of Nauel Huapi, only a little more than 1 percent of Argentina is given over to the country's thirty-three national parks, contiguous reserves, and strict nature reserves. They cover some 35,000 square kilometers (13,510 square miles), which is hardly more than a single park in some comparable nations. The leadership of Francisco P. Moreno, who gave the original land for Nauel Huapi, and that of Ezequiel Bustillo, Angel Gallardo, and Luis Ortiz Basualdo in the 1930s has been verbally honored but only weakly followed.[8] The largest nongovernmental conservation organization is Fundación Vida Silvestre Argentina (Argentine Wildlife Foundation), stimulated by the late José Maria Gallardo and engineered by Miguel Reynal, its president during its formative years, and Teodosio Brea and Francisco Erize. Although the foundation has a membership of less than four thousand, it has played a major role in bringing broad conservation issues to public notice and winning a place for environmental education in the school system. Among the oldest NGOs is the small (nine-hundred-member) but increasingly effective Aves Argentina, founded in 1916. As its name implies, it focuses on birds and has identified some four hundred Important Bird Areas (IBAs) in Argentina. However, the most effective organization in Patagonia is Fundación Patagonia Natural, which, with the collaboration of local authorities and the help of WCS, is playing a major conservation role in the Southern Cone. In the 1960s, when Bob Goelet led us to the elephant seals on Peninsula Valdés, two of these three organizations did not exist, and almost no one spoke up for wildlife.

The south Atlantic fisheries were little developed in the 1960s, and Patagonian coastal towns, outside the oil port of Comodoro Rivadavia, were very small. By the 1970s, hydroelectric dams in the foothills of the Andes brought power to the coast, a large aluminum refining and production plant called Aluar was installed in the beautifully positioned little town of Puerto Madryn on the broad Golfo Nuevo, and a major fish processing industry began to be developed.

Aluminum processing requires large amounts of aluminum ore and prodigious amounts of electricity. There are neither significant deposits of aluminum ore in Argentina nor significant electrical resources in the coastal desert. But ore could be shipped from New Guinea (yes, New Guinea), and rivers and lakes in an Argentine national park in the

Andes could be dammed and made to generate electricity, which could be relayed all the way across the Patagonian desert—if a little help could be provided to private investors at public expense.

Few environments are unsuitable enough, or are too beautiful, or have development costs too high to withstand the blandishments of subsidized entrepreneurs, as the United States' mismanagement of timberlands such as the Tongass and conflicts over the Arctic National Wildlife Refuge make clear—to say nothing of the perverse subsidies of its superfarms. Alejandro Lanusse, then president and commander of the Argentine army, is said to have taken "a personal interest" in Aluar, the aluminum plant on the shores of Golfo Nuevo. Aluar is now the area's largest single employer, its plant but a stone's throw from Puerto Madryn's resort-like beaches, where bikinis blossom each austral summer and whales court offshore in winter.

But it is not aluminum that has the most immediate potential to destroy Patagonia's colonies of seals and sea lions, penguins and cormorants. It is the commercial fisheries, which accidentally net them and compete with them for food. In the five years prior to 1996, the Argentine catch doubled to more than 1,200,000 metric tons, mostly from Patagonian waters. It is now about 1,800,000 tons. One fish farm project after another has been tried and failed, and yet more are planned. Argentines eat little fish, and most of the catch is exported. Mink and fox farms—often dirty and badly managed—senselessly subsidized by the provinces and established without "due diligence," are another unfortunate development. Several such operations have allowed mink to escape, with the result that they are said to have become a scourge to native waterfowl. Multimillion-dollar fish processing factories have also been installed without appropriate research. In Camarones, a brand-new plant has been built almost within sight of another that had already failed. Both stand empty. It is as though the investing agencies are both deaf and blind.

From the standpoint of wildlife, one of the most environmentally worrying by-products of new industries is simply more people—more human mouths to feed, more waste to process, and more leisure time to use guns and 4x4s. The population of Puerto Madryn, now about one hundred thousand, is twenty times as large as when I first saw the little town in 1964. Although Puerto Madryn's growth is exceptional, most Patagonian towns are expanding, and people are moving off the land. On the other hand, just as quantities of people can be an indirect threat, they can also be monitors and protectors, and conservation has proved to be much more a preoccupation of city people than country people.

All over the world, people typically pick the same places to live that wild animals do, usually because of available food and water, so most cities and agricultural areas have developed in or near prime wildlife habitat. This makes protecting wild animals by setting aside the places where they are best adapted to live an increasingly rare option. Ultimately, as human population continues to invade wilderness areas, conservation must be based on a combination of compelling inspiration, persuasive scientific information, local involvement, and, especially, transparent use of public resources to prevent wildlife extinctions. Some of the illustrative importance of the Southern Cone comes from the facts that human population is still low and conservation canons are more easily revealed on a desert coast and in the stark environments of steppe and altiplano with big, charismatic creatures than in the extravagant congestion of the tropics. However, in either place, simple protection is not enough for animals to thrive.

In 1947, for instance, the United States' Florida Everglades National Park was established "to protect the finest assemblage of large wading birds on the continent," and hunting and encroachment were halted. Nevertheless, by 1989, populations of the big wading birds had declined by 90 percent. The great lesson: Protection and benevolent neglect do not save wildlife communities when their ecosystems are subject to modification by activities outside their borders. The Everglades' water sources were compromised by upstream water diversion, by chemicals from agriculture, and by adjacent residential use. Invasive foreign plants and animals further changed the reserve. Food for waterbirds vanished. Nesting areas protected from predators by water disappeared. Now a restoration is being attempted—for $7.8 billion.

Even without external complications, protected areas alter over time. Changes in plant communities as they compete, mature, and succeed each other, for example, affect species-carrying capacity (the maximum population of a species allowed by the environment). Protecting a penguin population at its colony on a Patagonian shore but not the ocean creatures that provide its food is like building a cowbarn in a parking lot and ignoring the need for grass—or protecting the Everglades Park borders but not its water and plants. Only very large reserves can provide sufficient variety and adequate space for large animals without increasingly zoo-like care. The immense U.S.-Canadian Glacier National and Waterton-Glacier International Peace Park in Montana and Alberta has yet to lose a single species; but few refuges are so grand, and climate change may threaten species in even the largest. A study of seven of the biggest national parks in western North America found that the populations of twenty-seven species of mammals had gone extinct in one or more of even those reserves.[9] Again, plant succession, species competition, disease, and local events that affect one species affect others, as well.

Argentina and Chile have created sizable parks along their border, like the U.S.-Canadian joint preserves. These serve important diplomatic and political purposes as well as environmental purposes and protect important watersheds. However, the impoverished Administración de Parques Nacionales manages many Argentine parks poorly, and few Argentine scientists have been encouraged to lend their expertise to help. Chile's Corporación Nacional Forestal (CONAF) is no better off.

The wildlife populations of much of Argentina (except those of elephant seals), and those elsewhere in the Southern Cone, are so fragmented, their numbers so small, that many are teetering on the edge of disappearance. The colonies of birds and mammals on the coast of Patagonia have won only a single national park, as mentioned earlier, and little enforced legislation. None of the coastal reserves now set up by forward-thinking Patagonian provinces can protect the essential wildlife food supply in the seas that border those colonies. As a result, not one reserve approaches a minimum viable size. Although there are some fine rangers with deep personal commitments to the environment and new training programs under way in Patagonia, none of the reserves are consistently managed by trained rangers. What would it take to remedy that situation, to better ensure the survival of Patagonia's singular wildlife for the future, to stimulate the idea of conservation as a public policy and bring home an understanding of the urgency?

In 1964, providing more and better information seemed the best answer, and that meant science. Creation of awareness through scientific investigations that could call attention to the unique nature, the attractiveness, and the unusual accessibility of Patagonia's wildlife—and to its precarious state—was an obvious need. It would be especially powerful if generated by Argentines themselves. In any case, one does not walk into a foreign country, announce that he or she is from the land of good intentions, and proceed to "do conservation." Nor is there any one formula with which to initiate successful conservation projects. But getting facts about the nature of Patagonia's wild animals—how many there were, how they lived, and what was happening to them—would be a good way to begin.[10] Scientific questions are usually attached to physical realities that others can corroborate, while conservation questions are often tied to slippery human beliefs, aspirations, and economies. Newcomers can rarely convince residents about what needs to be done in their own surroundings without both compelling new information and a long-term personal commitment. To make a difference, WCS would have to develop a program of investigatory science, help train local people, and, where possible, show the positive economic implications of conservation. The research would have to excite Patagonians about the nature of the wildlife itself, what it could mean to Patagonians, and help to form a vision or series of visions for its future. The resolution of conservation dilemmas is almost always more value-based than peso-based.

## A Patagonian Conservationist in the Campo

In 1979, on my sixth trip to Argentina, I met a person who was to prove crucial to my understanding of Argentina and to my conservation aspirations. Graham ("Guillermo") Harris grew up in Neuquén, in northwestern Patagonia, and was sent to an English boarding school in the province of Cordoba. Tall as a Tehuelche, he made occasional money by guiding wildlife tours and, somehow, became an accomplished wildlife painter. He has an exceptional knowledge of Argentine birds and great people skills, and it is impossible not to like him. He has not only an unrepentant sense of humor, but also the abilities to fix broken-down vehicles and to draw improbably accurate maps by hand. These may not be essential abilities in someone who wishes to undertake fieldwork, but, as a former colleague, the late William Beebe, instructed me, "Always look for the kind of man who can fix a Ford." Although born and raised in Patagonia, Graham speaks Oxford-Cambridge English. He is also disconcertingly fluent in elephant seal, guanaco, penguin, and goat. We met as he was finishing his veterinary degree, uncertain about his future, and nearly broke—the last two being outstanding qualifications for the association I had in mind.

Graham had come to realize that he was not truly interested in treating the infirmities of cows and chickens, horses and sheep, and that he cared deeply about Patagonia's wildlife. He knew that he wanted to work with wild animals and to paint, which is an easy way to starve. His idealism lifted my spirits. WCS needed someone to keep an eye on its Peninsula Valdés whale camp, and I wanted a sensitive and informed ear there. Graham would prove indispensable in coastal work in Patagonia and, eventually, on the steppe and altiplano. He became the first of a gradually growing cadre of Argentine colleagues who have drawn most of the details in the picture of Southern Cone wildlife that has emerged.

Graham Harris's assignment to oversee the camp on Peninsula Valdés was both idyllic and uncomfortable. The station is only a few miles from the site of the ill-fated friars' church, on the edge of Golfo San José, a magnificent bay that takes a large windy bite out of the north side of Peninsula Valdés. It is no Walden Pond, and, unlike Thoreau, Graham could not amble to town for supplies or leave his laundry with his mother; there was, in fact, no help within walking distance. At the outset, he was by himself in that remote and stormy land, and, as the Argentines say, "more lonely than Adam on Mother's Day." Camp, as it is sometimes called, is an uninsulated little building of concrete block with a corrugated metal roof, secondhand furniture, a tiny stove, and an outhouse at a distance that, in Patagonian winters, encourages continence. The camp is only twenty or thirty yards from the golfo, and from passing sea lions and right whales that roll, splash, and sigh in the nearby surf; it shares its patch with guanacos, cuis, and peludos. The kitchen is adorned with a single cold water spigot, which trickles water from a tank supplied by infrequent rains piped from the roof or a water truck from a source over fifty miles away. This is, after all, in a desert. Sometimes leaf-eared mice, lizards, and snakes are found decomposing in the water tank. The camp is also equipped with a pickup truck, which has ranged in condition from terminal decrepitude to "pretty good."

The north end of the station is a workshop now filled with odd tools, life jackets, pumps, tires, Tehuelche scrapers and arrowhead chips, miscellaneous feathers, animal skeletons, and a workbench. Behind it is a tiny studio where Graham paints, when he can find time, and to one side is a closet-sized photographic darkroom. Outside the station proper are two small huts atop nearby cliffs that afford marvelous views of the golfo and the chance to locate whales, guanacos, seals, and whatever else happens to pass by. Below camp, near the water, is another structure housing the station's Zodiacs, those small, almost unsinkable, inflatable boats used where portability is essential.

About one hundred feet south of the station is a Quonset hut of corrugated fiberglass and concrete block that resembles an oversized armadillo and is called the Peludo Palace. This is where visiting scientists usually stay and where Kix and I bunk when at the camp. It is not cluttered with electricity, temperature controls, or ventilation—but parking is free. Seventy feet farther south, on a little hillock, is the infamous "Tehuelche's Revenge," the outhouse, which, when the door is open, provides one of the most magnificent coastal views on the golfo; when the wind is blowing, it whistles, wobbles, and could do with a seat belt.

After Graham accepted the WCS offer to run the camp, Patricia Franklin, whose dark-haired, gray-eyed beauty conceals a businesslike approach to living in the campo, gave up a promising career in a large Buenos Aires firm to marry him. Together, they set about making the camp as livable as possible and addressed the special task of writing and illustrating *A Guide to the Birds and Mammals of Coastal Patagonia*.[11] In time, they had a family. Sometimes, unwelcome neighbors made an appearance—Graham described the following incident in an article about life at the camp:

> "Mummy, Mummy! Come and see! Come and see!" The cries became persistent so Patricia looked up to see what our two year old daughter wanted. A shiver of horror ran up her spine and she caught her breath. Sabrina was sitting on her haunches, her little body leaning forward as she gazed at a pit viper (*Bothrops ammodytoides*) lying at her feet. Her eagerly pointing finger came within an inch of the snake's head.
>
> It was early November and the warm afternoon sun of the austral spring had brought the reptile out of hibernation, searching for mice and lizards upon which to feed. Quietly, with barely a trace of urgency in her voice, Patricia ordered Sabrina not to touch the snake and to step back. Puzzled, the child obeyed.
>
> Bringing up Sabrina and Edward on the wild and desolate Patagonian coast is an unusual experience. We don't worry about busy streets, kidnappings, or drugs but we do have to watch out for pit vipers in summer.[12]

Foxes, skunks, Darwin's rheas, partridge-like crested tinamous, occasional burrowing parrots, a nearby nesting black-chested buzzard eagle, even a visiting flamingo or, on very windy days, a huge black-browed albatross might be in sight. Patagonian mockingbirds and morning sierra finches are station regulars, and so are American oystercatchers nesting on the beach. Giant petrels, especially the big males looking for a dead seal or whale, are also regular visitors, and sea lions have a favorite resting spot at the foot of one of the nearby cliffs.

Even black-necked swans, steamer ducks, and red-footed gray dolphin gulls may appear.

The camp was created for studying right whales and dolphins. It was originally used by our imaginative former staff member and cetologist Roger Payne, and later by a long list of scientists who expanded Roger's studies of the whales and began new ones on dolphins, guanacos, and other creatures. The camp is now also used for training budding Argentine biologists, but one of its most important roles has been to serve as a center of commitment to and concern for the surrounding shore and bay. For nearly thirty years, it has functioned as a focal point of local wildlife intelligence, its visiting scientists acting as monitors and conservation ombudsmen, speaking out against overhunting, overfishing, overdevelopment, and other environmental folly — especially with Graham's quiet but authoritative voice.

No biologist knows that part of Patagonia or the station as well or has spent so much time there as Graham, Pat, and their children. "At first [living here] was not easy," Graham wrote, and continued:

> There were many difficulties to overcome, largely due to isolation . . . but we are not alone at the station. Right whales pass the winter and spring of each year in our sheltered bay, where the females bear and nurse their calves. On a clear day in September, as many as a hundred of these great beasts can be seen from the cliffs above the house. . . . As we lie in bed on calm nights, the sounds they make — blowing, and slapping the water with their flukes — are so loud and clear they seem to come from underneath the bed. . . . Flocks of shorebirds and an occasional flamingo land and stay until the tide comes in again.
>
> One day, walking along the beach with Edward, then two, we found our progress constantly slowed as he made frequent stops to inspect a shell or a piece of seaweed. . . . When he stumbled upon the first penguin carcass (I have counted as many as 80 along a half mile of this beach, some obviously the victims of oil pollution but many dead from unknowable causes), his curiosity was aroused. Puzzled why it lay so still, he demanded an explanation. Like so many parents faced for the first time with the formidable task of trying to explain death to a child who can barely talk, we balked. "It's asleep," we quickly assured him, thinking it was an easy answer for all of us.

> Later that day, Patricia met our son coming out of the bedroom. Two stinking penguin carcasses had been carefully tucked into our bed, their rotted heads tenderly laid on our pillows and the blankets pulled up under their chins. "Shh!" said a little voice earnestly. "They're asleep."[13]

As visitors, Kix and I have especially enjoyed the camp's hairy armadillos, or peludos (*Chaetophractus villosus*). They have adopted the Harrises, who maintain a kind of open-door benevolence toward all sorts of creatures. Tufted tit-tyrants nest on the porch, guanacos raid the struggling flower garden, guinea pig–like cuis, rarely seen in the bush, scamper back and forth in front of the porch, big leaf-eared mice have to be shooed from our sleeping quarters, morning sierra finches sing from the zampa bushes—and peludos walk right into the kitchen—with unerring precision. Their super noses lead them to food from surprising distances, and to garbage.

Whenever Graham could catch a peludo (often with its head in a bucket of refuse), he put a swatch of paint on its backside so as to tell them apart. Thus, on a recent visit, there was White Bottom, Blue Bottom, and Yellow Back—and sometimes a rather colorful Graham himself from various accidents as he pursued his spirited quarry through the bush with dripping paint bucket in hand. At one point, he attained a marked population of fifteen, and the result was that individual characters began to emerge: bully, clown, early bird, and so forth:

> One female, White Bottom, regularly appears at dusk and knocks at the back door, scratching with her long nails. When the door is opened, she trots in confidently and begins to search for fresh fruit and vegetables, which we put up out of reach. When she finds nothing in the kitchen, she moves to the living room and remains under the table in the hopes that the children will provide her with some of their dinner. Until satisfied that there is nothing edible for her, she refuses to leave. Although armadillos are not cuddly—all claws and armor—White Bottom is one of our children's favorite animal friends, and whenever they see her coming up the path toward the house, they give her a warm, noisy, welcome.[14]

Scattered far apart in the dusty, gray-green, and brown landscapes of Peninsula Valdés are a few substantial houses in addition to the usual simple homes no more elaborate than camp, and there are many tiny huts. Almost all of Graham and Pat's human neighbors in the campo work as sheep ranchers. Many are *puesteros* living at small posts or *puestos* with only their sheep dogs and horses. It is an isolated life in facilities that rarely have electricity or other conveniences, a lonely horseback and pickup truck community. A visit to the closest neighbors requires a lengthy ride or drive. Most of each day's work consists of maintaining fences, windmills, wells, and water troughs and looking after the sheep. Sheep seem designed to get into trouble, getting caught in fences, even in bushes, becoming separated from their flocks or their lambs, developing foot or eye trouble, and much more.

But then there are the cooperative efforts: the sheep roundups, shearings, disease treatments, and, finally, the special celebratory occasions with family and friends at the famous Patagonian *asados*. These lamb "cookouts" are laced with Argentine tradition and good fellowship, and always with the drinking of that strong, gritty tea, *mate*. It is enjoyed like teas and coffees elsewhere but much more intimately, for everyone sips from the *calabaza* (teapot) with the same *bombilla* (silver straw), which is passed wet and gleaming from mouth to mouth. Given the condition of the teeth and gums of some of the shepherds with whom one is likely to be sharing the bombilla, I always admire my Argentine colleagues' ready participation. For me, the practice requires a rigorous suspension of apprehension.

The nearly two thousand miles of coast from Peninsula Valdés to the Straits of Magellan, the focus of Graham's guidebook, is especially critical habitat in eastern Patagonia. That is where the coastal penguins, sea lions, and elephant seals breed and where most wildlife tourism is developing. It is also where the thirteen major towns are to be found, at least until one enters the Andean foothills. Changes there and in the sea itself will determine the future of all the seaside wildlife dependent for food upon the rich communities of invisible sea creatures living in the turbulent Atlantic to the east, which is being altered by drastic overfishing. That transformation, also affecting squid and whales, toothfish and albatrosses, endangers the very foundation of coastal wildlife. West from the coast, on the vast windswept steppes that stretch all the way to the Andes, guanacos, rheas, and maras are just as urgently and fundamentally threatened by overhunting, overgrazing by sheep, the introduction of invasive species, and predator imbalances, and are thus the subject of other WCS Southern Cone projects. In the Andes themselves, the world's most beautiful and least-known flamingos hang on in a series of spectacular high-altitude animal communities in widely scattered lagunas threatened by mining and other developments. They are now the subject of an Argentine-Bolivian-Chilean collaborative conservation effort.

In wildlife protection, three questions usually stand out: What species are there? How many? Why aren't there more? All relate to environmental carrying capacity simply because animal populations grow in suitable habitat until something stops them. Thus, all assumptions about sustainability, possible "animal harvests," and the preservation of wildlife populations rest on the capability of a particular environment to provide sufficient food, water, space, and shelter to sustain a viable number of some assemblage of species—and on how that species is constrained by humans or other species. Graham had to illustrate 185 birds and 61 mammals for his guide, and, numerous as that may seem, they are only a remnant, a small portion of dozens of mammals and birds, mostly large, that we suspect were "constrained" by the Foot Indians and their descendants, as well as by climatic and other changes, in the last twelve thousand years or so. The story of Patagonia's wildlife-human contact begins with Act I.

# ACT I

## *12,000 Years in Patagonia*

*For a transitory enchanted moment*
*man must have held his breath in the presence*
*of this continent, compelled to an aesthetic*
*contemplation he neither understood nor desired . . .*
F. Scott Fitzgerald

---

WHILE Paleolithic artists were sketching mammoths and rhinos on the walls of French caves, the first people in the Southern Cone were hunting strange-looking horses, giant sloths, and ancient guanacos. Everywhere I hike, on seashore, steppe, and puna, I find arrowheads, spear points, and the beautifully finished *boleadoras* (stone balls, formerly encased in leather and connected by long thongs) of the earliest Indians and those who followed them. One cannot but wonder about the humans who originally found their way into these forbidding latitudes. What happened to them? What do we know of their association with wildlife? In Patagonia's long drama of human-wildlife relationships, theirs was the first.

Seeking more food perhaps, or a better place, the first people wandered down into what is now Patagonia at least twelve thousand years ago. (The oldest evidence of humans in a nearby area is at Monte Verde in Chile, dated 14,700 BP. It is disputed, as are most claims of settlement for Patagonia prior to twelve thousand years ago.) They came in families and tribes, their ancestors having crossed the Bering Strait from Asia, each generation or so moving a little farther, each with only the most primitive of weapons and tools. It was an extraordinary trek, an adventure as great, as truly wonderful, as any in human experience. The climate was changing. It was a time of dramatic deglaciation and subsequent sea level change—rises of as much as sixty-five feet (twenty meters) in less than five hundred years.[1]

These peoples, called Foot Indians by archeologist Junius Bird, walked south through lands where no human creature had ever stepped.[2] They faced unknowable terrors and hardship but great abundance in game, and they tramped through vast landscapes that seemed without end. They carried their children and their few possessions in the presence of strange, sometimes huge, creatures. There were long-haired megatheriums—giant ground sloths as big as rhinoceroses—and smaller but similar mylodons, some of which survived until eight thousand years ago; saber-toothed tigers; glyptodonts, fifteen-hundred-pound, shell-covered, armadillo-like creatures nearly seventeen feet (five meters) long; huge ancient camels; peculiar horses (which paleontologists call hippidions and onohippidions); and rheas and guanacos like those of today. The migrants ran from those creatures and hunted them.[3] They walked in one direction and then another, and their tribes came in waves, sometimes just ripples, but a few were always headed south. They came by generations, not by destinations, making love, giving birth, fighting, and dying.

They hunted and foraged all the way from Alaska to the Straits of Magellan, and, while the waters were still low, some crossed the Straits into Tierra del Fuego. The few clues they left behind suggest that the trip took only a thousand years or so. How could they have gone so far, so fast, on foot through the trackless wilderness? To doubters, geographer Jared Diamond notes that an eight-thousand-mile expansion of people to Patagonia from, say, the U.S.-Canada border, in a thousand years, is only eight miles per year, "a trivial feat for a hunter-gatherer likely to cover that distance even within a single day's normal foraging."[4] It wasn't quite like that, of course. Those first Patagonians-to-be would have traveled many times eight thousand miles, meandering here and there, wherever prospects for food looked best and where ice and snow permitted, for the glaciers of the last ice age still covered much of North America between twenty-five thousand and twelve thousand years ago. As hunter-gatherers, however, these people moved far faster than the agriculturists who spread over western Asia three or four thousand years later at an average rate of less than one mile per year.[5] Like so many great human migrations, the trip was heroic only in retrospect. Day to day, it was a struggle to survive.

Unlike humans, horses and camels evolved in North America, and they preceded people to Patagonia sometime after the first isthmuses joining North and South America arose, about 2.8 million years ago. Skunks and peccaries also preceded *Homo sapiens*, and mastodons, wild dogs, foxes, bears, cats, raccoons, tapirs, and deer all made the trip. Most were there when the Foot Indians arrived, as were the rhino-sized ground sloths, giant camels, and huge armadillos, all of which, including the horses, disappeared soon after human arrival—although ground sloths survived in Cuba until just sixty-two hundred years ago.[6]

Archeological studies confirm that these pre-Tehuelches killed mylodons and hippidions and also megatheriums, glyptodonts, and onohippidions. They almost certainly played a role in the complete disappearance of those creatures, as their forebears had in the disappearance of mastodons and giant bison from North America some thirteen thousand years ago.[7] This explanation is controversial, however. Some paleontologists argue that climate change, especially that last ice age, was the cause of a great demise of megacreatures that took place around the world. Others argue that big extinctions took place at different times in different places, not in accord with any known climate change but often closely correlated with the arrival of human bands. In any event, forty-six genera, chiefly of large animals, are known to have gone extinct in South America in the past fifteen thousand years, most by ten thousand years ago—and most soon after the arrival of humans.[8] Although those creatures differed in many ways, they all tended to be either likely prey or likely competitors of humans.

Thus, the early Patagonians are accused, if not convicted, of having behaved as humans have behaved almost everywhere else—as the Maoris did much later when they killed off the great ostrich-like moas in New Zealand, for example; as the Indonesian immigrants did with Madagascar's giant lemurs and elephant birds; as the Australian aborigines may have done with carnivorous kangaroos, tent-sized tortoises, and giant monitor lizards thirty-five thousand years earlier; and as the Hawaiians did with nearly one hundred species of endemic birds well before the advent of Western settlers. Whatever the facts in Patagonia, the proto-Tehuelches ate or used almost everything at hand, even the body shells of giant armadillos, which they employed as roofs and tombs.[9] The animal supply may have seemed endless.

Shorelines and shallows must have been filled with shellfish; every rocky punta covered with trusting sea lion, seal, or seabird colonies; and the vast plains with immense herds of guanacos, rheas, and their antecedents. I wonder how long the chance to kill easily, that "ecological release," as it is called, persisted? Hundreds of years? Thousands? Did it enable the Foot Indians to become, for some golden age of abundance, far more numerous than the scattered peoples, such as the Tehuelches, Westerners discovered some 480 years ago?

Populations of most creatures, human hunter-gatherers not excepted, are dictated by the worst of times, not the best. I suspect that any golden period of numerous Foot Indians fed by uncontrolled access to abundant naive animals, as has been suggested for some other peoples and places, was short-lived if it occurred at all. There is no archeological evidence of large ancient human populations. Patagonia was, even then, a tough place for people to live, a cool semidesert where the Andes had cut off the rain.

Long before, 12 million years ago, the gradual elevation of the Andes was already significant though modest. Then, between 4.5 million and 2.5 million years ago, there was an extraordinary rise of between six thousand and thirteen thousand feet. (The Andes are still rising about half an inch a year.) The result was interception of the wet winds off the Pacific and the creation of a vast rain shadow throughout the eastern steppes. Today, only a few miles east from the mountains, precipitation drops by an order of magnitude. A little less rain, and the steppes would be not a semidesert but a wasteland of sand and rocks. A little more, and it would be a forest. Whatever the effects of the new Patagonian people on the wildlife, the growing desert they found must have been a better place for camels than for ground sloths.

Eventually, the Foot Indians of the Southern Cone split into subgroups, tribes called Araucanians and Puelches, who lived north and west of Patagonia, and Onas, Yaghans, and Hausch, who lived south in Tierra del Fuego—and the aforementioned Tehuelches.[10] These last were a handsome tribe of unusually big people who chose to live nomadic lives on Patagonia's harsh steppes and blustery coast between the Rio Negro–Rio Colorado area in the north and the Straits of Magellan in the south. They were, above all, camel hunters, and it is in the dry soils of that area that I have often found their arrowheads and boleadoras.

Patagonia's guanaco is a camel, a pony-sized, humpless camel that flourished in the aridity of the steppes. Its herds reached the tens of millions, maybe as many as forty million by the time Europeans first arrived. The rhea prospered, too, and, with the guanaco, became the backbone of the Tehuelche's simple hunting-gathering economy, of their religion and food and clothing, and even the materials for the crude tentlike *toldos* in which they lived. The lack of a more complex culture and economy may reflect the small size and scattered nature of Tehuelche populations. Numbers are needed to build cultural creativity and economic complexity. Although intriguing paintings and engravings of uncertain meaning are found in caves and on sheltered stone formations here and there in Patagonia, most apparently predate the Tehuelches. Some feature hundreds of human hands, while others are scenes of humans surrounding and attacking running herds of guanacos. At any rate, the effect of these peoples on the species of wildlife still extant today must have been modest, whatever that of their ancestors on the creatures of twelve thousand years ago. Even into the twentieth century, there were teeming colonies of sea lions, seals, and seabirds breeding on the Atlantic beaches and headlands. There simply weren't many Tehuelches to hunt them—no matter how large those hunters were.

The Tehuelches' stature was famously magnified by the reports of Ferdinand Magellan and his crew. Besides being the first to circumnavigate the world, the expedition's members were the first Europeans to visit Patagonia. With some courage, they spent more than six months in the austral winter of 1520 repairing their boats along the coast, at a place Magellan named San Julián. Four years before, Juan Diaz de Solís, who was one of the first European explorers of northern Argentina and discovered the great Rio de la Plata, was reported to have been killed and eaten by cannibals near where Buenos Aires is now.

Magellan's expedition branded Patagonia "a land of giants," claiming that crew members stood barely as high as the Tehuelches' waists. Darwin's observations provided a different picture:

> We had an interview . . . with the famous so-called gigantic Patagonians, who gave us a cordial reception. Their height appears greater than it really is, from their large guanaco mantles, their long flowing hair, and general figure: on average their height is about six feet, with some men taller and few shorter; and the women are also tall; altogether they are certainly the tallest race which we anywhere saw.[11]

It was Magellan's diarist, Antonio Pigafetta, Knight of Rhodes, former assistant to His Excellency the Roman Ambassador to the Court of King Charles I of Spain, whose chronicle gave rise to the name "Patagones," thus "Patagonia." The word has been doubtfully interpreted as "large footed." After all, their feet were swathed in guanaco skins. However, historian Ramón Lista claims, rather indignantly, that the word is from a Quechua-derived Tehuelche term, *patak*, for "hundreds," as in, "We are hundreds."[12] Writer Bruce Chatwin came up with a more convoluted explanation, suggesting that Magellan adopted the name of a monstrous beast from a book published by an unknown author in Spain in 1512. At the moment, further scholarship seems unlikely to shed more light.

While in San Julián, Magellan dealt summarily with a mutiny in his little fleet. Luis de Mendoza, leader of the mutiny and captain of the *Vittoria* was killed and his body drawn and quartered. The same fate was visited upon Gaspar Quesada, captain of the *Concepción*. Magellan left one of the mutineers spitted on the shore, while two others, Juan de Cartegna and Father Pedro Sanchez de la Rena, were sentenced to be marooned and were put ashore two months after Magellan left

San Julián. For the Tehuelches, these leavings could not have been reassuring.

Subsequent European visits to Patagonia brought settlers. In 1539, there was Francisco de Camargo, who left an exploratory expedition of 150 men in Patagonia—nothing is known of their fate, nor has any trace of them ever been discovered. Nevertheless, rumors that they had found great wealth led to at least four expeditions to search for them and many hopeful legends, which lasted well into the nineteenth century. It was said that the men set up the colony of "Traplanda," described as a community of white Indians and sometimes called the enchanted "City of the Caesars," which has not yet been found.[13]

Sir Francis Drake followed Magellan at Port San Julián in 1578 and, with dismaying coincidence, also put to death two and marooned another of his captains who had mutinied. Drake's chaplain added to Pigafetta's description of the Tehuelches: "They don't cut their hair but keep all manner of things in it. They carry a quiver for arrows, knives, a case for tooth picks, and a box for firesticks." They have "clean, comely and strong bodies" and are "very active, a goodly and lively people . . . fond of dancing with rattles around their waists."[14] Later, near the eastern entrance of the Straits of Magellan, Drake's crew killed "no lesse than 3,000" penguins on Santa Magdalena Island for provisions. Subsequently, the island became a regular provisioning stop, and, in 1594, Sir Richard Hawkins commended the unlucky penguins, observing that "they are reasonable meate rosted, baked or sodden; but best rosted."[15] Elizabethan sailors are said to have believed penguins could be the souls of their drowned companions—but that did not save them from the pot.

The year after Drake's visit, 1579, Pedro Sarmiento de Gamboa was dispatched from Callao to search Magellan's Straits for Drake, the Spanish fearing that the famous seaman was attempting to lay claim to Patagonian lands. (They were right to be suspicious. Drake had made a fortune preying on Spanish treasure ships. From one voyage, he returned to England with spoils worth, in today's money, $60 million.) Amazingly, Sarmiento saw Tehuelches who chased their game on horseback and brought the animals down with boleadoras. It was less than forty years since horses were first imported by the Spanish nearly a thousand miles north at the Rio de la Plata. By Drake's time, there were already thousands of horses roaming the pampas; many seem to have resulted from the five mares and seven stallions abandoned by Captain Pedro de Mendoza in 1541, when Indians isolated the town he had set up near present-day Buenos Aires and starved the settlers out.[16]

In 1581, Sarmiento was sent out from Spain again, this time with twenty-five hundred people in twenty-three ships, to found new colonies in the Straits, Spain still being suspicious of the colonial aims of the British. Sarmiento did establish a settlement, leaving four hundred men and thirty women with eight months of provisions. But on the way home, the English captured his boat, and the colonists were forgotten. Five years later, Thomas Cavendish arrived at the settlement and found only twelve men (some say fifteen) and three women surviving.[17] According to some reports, two hundred of the men had been turned out to shift for themselves because of lack of supplies, and those two hundred, not those put ashore by Camargo, were the people who established the legendary city of Traplanda. In any event, most of Sarmiento's colonists who remained at the colony died of slow starvation and disease. Cavendish named the place Port Famine.

Patagonia is no place for the uninitiated. The Tehuelches could find enough to eat, however, and, in fairness, settlers came to sad ends in many places. Of the six thousand settlers sent by the London-based Virginia Company to Jamestown, Virginia, between 1607 and 1625, forty-eight hundred quickly perished.[18] Still, Patagonia is undeniably forbidding. It was one of the last areas in the world's temperate climate zones to be settled by Europeans and the very last part of Argentina, partly because of its unfriendly soils and climate and partly because of its unfriendly stone age Indians, who had not discovered the wheel, had no written language, and grew no crops. Although their numbers were small, they could be fierce, and they had no use for the settled ways or products of the Europeans, except for their horses.

Horses brought the Tehuelches an entirely new way of life, and they quickly became a skilled and mobile cavalry, very like the Sioux and

Blackfoot would later become in the U.S. West and just as formidable. Their preferred war weapon seems to have been a spearlike lance as long as eighteen feet, though, in later years, a few managed to obtain firearms. They employed both against some who settled on their land, burning their estancias to the ground and killing everyone they could find, but maintained cordial relations with others. For many years, they prevented settlement on Peninsula Valdés, where the friars were killed in 1810.

North of Patagonia, on the vast, fertile pampas, Europeans gradually extended their hold, building a rudimentary economy based, at first (1600–1750), on hunting feral horses and also cattle, some of whose ancestors had been turned loose as early as the 1570s, for their hides; then (1700–1850) on the gradual formation of herds of partially tamed animals; and, finally, on extensive cattle and sheep raising.[19] Consequently, in the late 1870s, military campaigns were directed at the pampas Indians, and those who survived were quickly pushed below the Rio Negro. Finally, these "civilizing" activities began to move south into Patagonia, and into the dry, sparsely peopled Tehuelche lands. Most of what we know of the Tehuelches by this time is based on the word of one extraordinary man.

Commander George Chatworth Musters, late of the British navy, joined a band of Tehuelches near Punta Arenas on the Straits of Magellan in April of 1869 and traveled alone with those people for more than a year. He rode and walked with them over the steppes and deserts, through incessant winds and frequent storms, all the way from the Straits across what is now the huge and still largely uninhabited province of Santa Cruz, north to the Atlantic coast, where Rio Santa Cruz enters the ocean, and then west to the foothills of the Andes, north to the Rio Negro, and east to the town of Carmen de Patagones, near the coast. In that amazing trip, he traveled through the lands of both the southern and northern Tehuelches and up to those of the Araucanians. He left a wonderful account, *At Home with the Patagonians*, filled with descriptions of Tehuelche customs, lifestyle, internal battles, and behavior as a people, for whom he developed great respect and affection.[20] Among other things, he tells us that, in contrast to many early societies, the Tehuelche men took only one wife at a time and were tender and respectful toward their women. Musters noted that they were superb horsemen and that he never saw an Indian horse that was not perfectly quiet and "biddable." Even the smallest child could safely ride his father's warhorse. By that time, the Tehuelches were clearly using the boleadoras and long lances in preference to the bow and arrow.

The dry steppe offers little that any large number of people can eat, little place to plant crops, and little to support large resident, as opposed to small nomadic, populations. Imagine the life of the Indians when they had no transportation but their feet. Except along the rare watercourses, there were no trees, just huge, petrified araucaria trunks from a genus so ancient it had provided shade to dinosaurs[21] and relentless wind, wiry grass, thorny bushes, endless vistas of semidesert and sky—and quick-footed guanaco herds and rhea flocks. Hill-like coastal middens of mussel shells still bear testimony to the ages over which the Indians scoured the seashore for food, and to its abundance there. Some middens in Tierra del Fuego are more than two thousand years old. As the Europeans appeared, many of the hardy Tehuelches—of their few—were felled by smallpox and measles and probably by colds and flu, accidentally introduced by the new people who began to claim the land, a repetition of the fate of the Aztecs and Incas to the north in the wake of Spanish contact and conquest.

In the end, the majority of Tehuelches not killed by disease were slaughtered. General Julio A. Roca led a military expedition into Patagonia in 1879, nine years after Musters left his Tehuelche companions near Carmen de Patagones and forty-six years after Charles Darwin's visits there on the *Beagle*. In one extraordinarily murderous drive, with six thousand men, he exterminated most of the remaining pampas Indians and Tehuelches and opened the wild Patagonian campo to settlement and the establishment of estancias—and sheep. On January 1, 1885, the last of the Indians surrendered to the Argentine government, ending all resistance to nonindigenous claims and invasions. In the same year, many of the soldiers who had participated in the "Conquest of the Desert" were awarded millions of acres of land in Patagonia, continuing and expanding the Argentine tradition that supports very large landholders. Thus came to an end 330 years of Indian attempts to defend their homelands and ways from different peoples as a new Argentine nation took shape, eighth largest in the world in geography. Just five years later, on December 29, 1890, the last major battle of the North American Indian Wars took place, the Battle of Wounded Knee, at which some two hundred Sioux men, women, and children were massacred by the U.S. 7th Calvary.

The more I learn of the Tehuelches, and with every stone arrowhead and boleadoras I find, the more I regret the disappearance of those strong people so admired by Musters and others who knew them well. The lore of the proud Tehuelche is nearly absent from most of Patagonia now, except for kitchy souvenir trade items and collections of arrowheads, scrapers, stone knives, and boleadoras. Those last show up most commonly mounted on the walls of little stores and gas stations and represent their owner's personal treasures. On the road from the coast across the desert steppe to Esquel in the mountains, there is such a station, where, for several years, a pretty dark-skinned little girl has hung about, bashfully twisting her fingers and thin little dress in embarrassment, trying to sell finely finished arrowheads to strangers. Her mother makes them.

Darwin relates a poignant tale of three young Indians, perhaps Tehuelches, captured by the troops of the former governor of Buenos Aires, General Juan Manuel de Rosas, four decades before the final struggle with Roca:

> In the battle four men ran away together. They were pursued, one was killed and the other three were taken alive. . . . They were remarkably fine men, very fair, above six feet high and all under thirty years of age. The three survivors of course possessed very valuable information; and to extort this they were placed in a line. The first two being questioned, answered, "No sé" (I do not know), and were one after the other shot. The third also said, "No sé"; adding, "Fire, I am a man, and can die!" Not one syllable would they breathe to injure the united cause of their country.[22]

# ACT I

A new stream of immigrants had come to Patagonia, not on foot, not with bows, spears, or boleadoras, but also seeking a better life. Act II in Patagonia was beginning. The new tribes were from Spain and England, Italy and Germany—not a flood but a steady stream of immigrants, some soon recruited by the sheep industry and some in occupations more violent.

# ACT II

## *Hunting's High Tide*

*The hunting began as the sealers crept up the flanks*
*of the colony, taking advantage of wide beaches at low tide,*
*cutting off the animals from the sea. The sea lions lunged forward*
*to the sea but they were stopped by the sealers, who then hit them*
*with sticks, bars and, occasionally, fired at them with rifles.*
*The dead or unconscious animals blocked the escape of those*
*coming behind and these soon suffered the same fate.*
Claudio Campagna and Luis Cappozzo

---

HUNTING, constant hunting. To its new Euro-Argentine proprietors, Patagonia may have seemed almost as rich in wildlife as it did to the Foot Indians twelve thousand years before. In the 1800s, there were literally millions of guanacos and rheas, and the shores were covered with seabirds, seals, and sea lions. This time, however, the nature of the hunters, their weapons, and their purpose were very different. The killing of a half dozen guanacos or sea lions to meet immediate needs was no longer typical. Now market hunters were killing, not for subsistence but for export to large national and international markets; this was the first intimation of the power that the globalization of trade would increasingly exert upon Patagonia's wildlife. These market hunters wanted thousands upon thousands of hides, and they had middlemen, transport, and guns.

To kill rheas, they didn't even need guns. The feathers of those big ostrich-like birds were a major part of the Patagonian economy in the second half of the nineteenth century, and it has only recently come to light that that occurred through an overlooked commerce between the "fierce" Tehuelches and Welsh colonists in Chubut, the provincial home of Peninsula Valdés.[1] In fact, in the 1860s, the sale of vast quantities of rhea feathers was the most important export of the Welsh colony, its dollar value greatly exceeding that of wheat and sixteen times as valuable by weight. The trade was made possible by *un gran comercio de pieles de guanaco y pluma de avestruz* ("a big traffic in guanaco skins and rhea feathers") obtained during annual visits from the Tehuelches, who captured the rheas with their boleadoras and ranged far and wide on horseback to do so, even as they themselves were being hunted by Argentine soldiers.

How many feathers constitutes "vast quantities"? Between 1862 and 1884, the annual amount from all Argentina ranged from 31,000 kilograms (68,200 pounds) to 73,000 kilograms (160,600 pounds). Scientist Marcelo Gavirati notes that one must kill 2 rheas to get just 1 pound of feathers and calculates that, during this period, about 264,000 rheas were killed for the trade every year. For the most part, the available data did not allow him to separate the Darwin's rheas of Patagonia from the greater rheas, which live north of Patagonia. Where he could make a distinction, using only the figures from the Patagonian Welsh colony as Darwin's rheas, he found that those constituted about 22 percent of the kill. He also found that the amount of feathers exported went into a steep decline after 1880. Presumably, that was the result of overhunting—although the war against the traders' own suppliers, the Indians, must have played a part. We will never know how many rheas existed before that trade got under way (we don't know how many there are now), but it was certainly millions. Even today, the hunt has not stopped.

Between 1976 and 1984, 204,322 skins and 7,745 kilograms (8.5 tons) of feathers of the greater rhea and Darwin's rhea were exported from Argentina.[2] Juan Carlos Godoy found that between 1956 and 1960, 22,295 kilograms (24.5 tons) of rhea feathers (representing over 98,000 birds, of which perhaps a quarter were Darwin's rheas) were exported for $115,662 (in current dollars), $5.19 per kilogram, primarily for feather dusters.[3] With the paving of dusty dirt roads, the demand for dusters has declined but not ceased. Such hunting is not new, though it was not always done with rifles. Sir Francis Drake's chaplain described the Indian method of stalking Darwin's rheas over seven hundred years ago: "They hide behind a plume of ostrich feathers on a long staff."

In the twentieth century, hunters had guanacos in their sights more than rheas. In 1916 alone, between 30,000 and 40,000 baby guanaco (*chulengo*) hides were exported just from Chubut. In 1928, 300,000 guanaco hides were exported from Argentina;[4] between 1972 and 1979, 443,655 adult guanaco skins were exported from Argentina; and between 1976 and 1979, 223,610 chulengo skins, valued at $5.6 million, were exported.[5] As a hunt, the guanaco kill was not nearly as deadly or as fast as that which destroyed North America's bison. Guanacos are not so easy to hunt, and the rugged landscape did not lend itself to the buffalo hunter's trainload-at-a-time kind of butchery. But, inevitably, the once great populations declined to reappear as bedspreads, blankets, and leather products.

For a guanaco or rhea, death by clubbing or shooting is noisy and fast. Death by sheep is quiet and slow. With the establishment of large-scale sheep farming in the late nineteenth century, the new Patagonians began to strangle the fundamental productivity of the land. The steppe's original vegetation varied from sparse desert shrub to sparse desert grass; "sparse" and "desert" being the key words.[6] Only a few areas could support cattle. Little could be grown, except in the foothills of the mountains and the rare river valleys. The soldiers who hunted the Indians from Patagonia were followed not by farmers sowing crops, or cowboys or gauchos with cattle, as in much of North America and the Argentine pampas to the north, but by sheepherders. With the defeat of the Tehuelches in 1885, the woolly herds' numbers reached a score of millions—and began eating away the thin green covering of the sere soil that makes up the grassy steppe savannas, thereby destroying the soil itself. Sheep, in effect, kill native grazers slowly by stripping the land of its vegetation and trampling the soil to powder with their sharp hooves. Ecologically, one might think that they were but replacing the guanacos, but they did so with a ruinous difference. The sheep had no history in that land, no relationship with the plants evolved in the give-and-take of the millennia, no behavioral traditions of territory, spacing, and migration that might have rationed their grazing and their numbers. And the shepherds chose to disregard the messages that increasing dust and bare soil had written on the landscape.

Just offshore, commercial whaling was under way well before the Tehuelches were gone. By the middle of the nineteenth century, over seven hundred whaling vessels were hunting off the coast of Argentina, and they killed nearly all the southern right whales. This was in no sense a "harvest"—it was a slaughter, a nearly complete annihilation of the southern right whale almost as deadly as that of the plains bison, and it made that fifty-ton marvel one of the world's rarest mammals. At the same time, commercial sealing had begun on the Patagonian shores, often by sailors from whaling vessels, much of it continuing deep into the twentieth century, and in Uruguay and Chile even today.

There were once scores of naive colonies of sea lions and fur seals along the coast. Get-rich-quick entrepreneurs employed jobless immigrants to encircle dense breeding colonies, where the animals were most vulnerable, and kill them for their skins and oil, soon turning the great shoreside *loberías* (sea lion colonies) and the few *elefanterías* (elephant seal colonies) into boneyards. Old photos show gangs of men, caps and boinas (berets) firmly in place, surrounding bawling sea lions, mostly females, and bashing them with heavy truncheons. One can only imagine the shouting, the anguished screams of the sea lions, the sickening thwacks of the clubs, the gore; it must have been horrendous. And then the skinning, sea lion hearts often still pumping. More than 500,000 Patagonian sea lions, some elephant seals, and unrecorded numbers of fur seals were killed in these grotesque orgies. At least 268,000 sea lions were killed on Peninsula Valdés alone, and the slaughter continued until 1960, just four years before my first visit there. At Punta Norte, 149,070 were clubbed. Now, less than 500 can be counted.[7] The United States, especially the port of Boston, was one of the largest markets for these skins and for big kills in the Falkland/Malvinas in the nineteenth century.

Meanwhile, as though dwelling in another world, in the closing decades of the nineteenth century, pioneering scientists were excavating extraordinary fossils of extinct animals from the same ravines and gullies where guanacos were being shot and the same beaches where sea lions were being slaughtered. It is a tale that revolves around two remarkable brothers, Florentino and Carlos Ameghino, born in 1854 and 1865. Although their father, an immigrant from Genoa in northwest Italy, was never far from financial emergency, the brothers became astonishingly productive paleontologists, steadily digging up the rich wildlife of Patagonia's past during the years 1887 to 1903. "Their achievement was one of the most remarkable in scientific history," American paleontologist George Gaylord Simpson commented.[8] However, Florentino's skill as a fossilizer wasn't finally matched by his judgment; he eventually persuaded himself that "all the mammals of the world have as their ultimate ancestors, Patagonian marsupials of various sorts (including man)."[9] (Fossil finding in Patagonia easily supersedes much else. In 1833, Darwin had written home, "There is now a bloody war of extermination against the Indians—so fine an opportunity for geology was not to be neglected, so that I determined to start at all hazards.")[10]

In 1904, the year after Carlos Ameghino's last expedition to Patagonia, former general, now president, Julio A. Roca accepted a donation of land in the Andes from Francisco P. Moreno in the magnificent Nauel Huapi area as Argentina's first national park. By 1922, the park had been enlarged from 7,500 hectares to 785,000 hectares (1,938,950 acres), and by 1937, four additional national parks—Lanin, Los Alerces, Perito Moreno, and Los Glaciares—had been gazetted.[11] Yet not one park was established on the coast. No attention was given to the disastrous slaughter of wildlife occurring there. The parks might as well have been founded for paleontologists (the same could have been said about the founding of Yellowstone in 1872 amid the destruction of the bison and the passenger pigeon). Parks were thought of in terms of magnificent landscapes, not magnificent wildlife, and that perception has not disappeared. I recall the disgusted observation of a Tanzanian parks official who visited me during a study tour of western U.S. national parks; he said, "I asked U.S. parks officials what they were preserving, and they showed me rocks."

# ACT II

And so, ignored by scientists, politicians, and citizens, the wildlife slaughter dragged on. The great Patagonian sea mammal population declined to near extinction. The seal, sea lion, and whale killers destroyed a natural wonder and a national resource—and put themselves out of business. The vast southern shores of Argentina fell silent.

---

The two-thousand-mile Argentine coastline is mostly high cliffs, shingle beaches, and polished pebbles, a landscape sculpted by strong winds and rough waters. The soils are yellowish, thin and dry, skinned of vegetation by the screaming winds of the "Roaring Forties" (the high winds in the 40° latitudes). They form the terrestrial edge of a vast, shallow, underwater ledge, the broadest continental shelf in the Southern Hemisphere. At its eastern verge, this huge submerged steppe drops steeply into the Atlantic abyss and is bordered there by a cold ocean river of Antarctic water.

This ocean river, the Falkland Current, spins off the Antarctic gyre, surges around the Falkland/Malvinas Archipelago 280 miles due east of the Straits of Magellan, and sends subflows west and north along the shore as the Patagonian Current. But its major flood continues northward at the margin of the continental shelf even as its streams and eddies wash the coast all the way up past the Rios Negro and Colorado at the top of Patagonia, where it meets the warm Brazil Current. It is rich in nutrients and plankton and acts as both highway and cafeteria for great schools of fishes and clouds of invertebrates, which, in their turn, attract pods of whales and dolphins; herds of sea lions, fur seals, and elephant seals; and flocks of albatrosses, petrels, penguins, and cormorants. They have foraged, interacted, and competed there for countless millennia.

The diversity of sea mammals and birds that breed along the coast of Patagonia is now small. There are three species of pinnipeds (fur seals, sea lions, and elephant seals), a half dozen or so cetaceans (whales, dolphins), and sixteen species of marine birds. But their biomass, their sheer weight, is great, and when the modest demands of the Tehuelches ceased, their numbers, hence their biomass, was probably much greater. The Patagonian continental shelf with its productive shallow water is so immense, the history of huge populations of marine mammals here so tangible in the landscape of eroded sea lion scapulas, cracked mandibles, broken teeth, and occasional whalebones, that, on dark foggy days, one senses their ghosts on the long, empty beaches and listens for their voices. Sifting through these bony bits and pieces, looking for clues to the age and sex of their owners, makes one reflective. What must the sea lion clubbers have been thinking? Even had they no mercy, surely it was clear that they were destroying the source of their jobs.

As the years passed, the few sea lions, fur seals, and elephant seals that had found refuge in out-of-the-way places began to grow in numbers, largely unnoticed. Gradually, the stage was set for their recolonization of the beaches. It was at that point, in the early 1960s, ignorant of most of the Southern Cone's history, that I came to Patagonia to count elephant seals and wonder at the flamingos, penguins, and rheas. The Patagonian steppes and shores had been empty of nomadic Indians for almost eight decades. Autos and trucks bumped along old Indian footpaths that had become dirt roads, shrouded in blinding clouds of choking dust. Occasional airplanes and the first satellites passed overhead, along with unseen radio and television waves. Food and material goods from the north were gradually, very gradually, becoming available. Wool, fishing, and oil supported a growing population. The features of human life had been transformed, but the face of much of the land and water was unchanged.

Guanacos and rheas still lived on the steppes, despite continuing overhunting and the thoughtless export of excessive numbers of skins and feathers. A tenuous rebirth of marine mammal populations could be seen along the coast, where its near total destruction had brought killing to a halt. Even a small herd of right whales had reappeared to court along the shores of Peninsula Valdés, while large colonies of Magellanic penguins still bred on the shores, as did imperial cormorants, rock cormorants, terns, rare dolphin gulls, and much more — and hardly anyone knew anything about them, or cared. Without sanctuaries and caring, all were liable to be lost again.

The Tehuelche's cold semidesert remains a strangely unsettling landscape of immense plateaus, dark cliffs, and arid cañadas, prehistory palpable in its petrified forests and fossil-strewn soils. Its hills and valleys are molded over sedimentary and volcanic rocks from the Mesozoic and Cenozoic ages, and its desolate basins are pockmarked with salty ponds and lakes. Endless gray-brown plains disappear in steep purple hills, their shapes wandering in dusty mirages beneath the hot sun. But for me, the allure of Patagonia is most compellingly rooted in its wildlife, a bizarre assemblage of creatures gradually being displaced by immigrants and entrepreneurs in the spirit of a new national frontier.

As my visits continued in the 1960s, I met a scattering of Argentine scientists and public officials with sympathetic interests in wildlife, but, in trying to turn our concerns into conservation action, we faced a Maginot Line of public ignorance and disinterest—which proved opportune. No one objected to the investigation of the curious lives of whales and camels, sea lions, and penguins. It was considered harmless. In many places, Act II's direct kinds of wildlife destruction had been winding down, inevitably, as the number of animal targets declined. Threats to their possible recovery were becoming more subtle and complicated. And so we became hopeful enough, or naive enough, to believe that there could be an Act III, that uncovering the stories of Patagonia's unique wildlife, species by species, would open the doors of local interest and even pride, and that science-based conservation could win a leading role.

# ACT III

## *The Road to Conservation*

*Interviewer: "Do you think that we have the right to be hopeful?"*
*Jorge Luis Borges: "No. We have the obligation to be hopeful."*
Jorge Luis Borges

---

TO OPEN a truck door against a Patagonian wind is to risk having it torn off. Windstorms treat the steppes as tempests do the shore, eroding the soil, polishing its rocks, and twisting its plants. Patagonians take pride in their wind, outdoing one another in descriptions of it to newcomers, in which, at the very least, gravel is blown through the air with disastrous effects for windshields and flying birds are blown backward—and they are. But Patagonia is not one place; it is many. You can be swallowed in wind and dust, or in big, beautiful, lung-filling, mind-swamping, unlimited space. You can drive it from side to side and from end to end and, unless you know where to look, see little wildlife, just endless sheep fences, plains, rocks, and sudden valleys.

Driving between always distant destinations, hour after hour against a 50-mph gale, dodging potholes on indifferently paved roads, or trying to ride the tire marks of the truck before yours on unpaved routes, veering from one side of the track to the other, can push you beyond fatigue into disorientation. The roads are notoriously bad, making you constantly apprehensive of the sudden lurch and bubbling rumble that heralds a flat tire—Graham and I had thirteen in ten days on one memorable trip—and the search for gas in wayside villages can rival the search for wildlife. Clocking more than 100,000 kilometers together on Patagonian roads, we have developed an hour-on, hour-off driving system, usually monitored by Kix. During Graham's shift, I scan the plains for wildlife, engage Graham in lengthy discussions of the political constraints in an Act III, or doze and think of glyptodonts.

The attitudes of most Southern Cone people toward wildlife, I found in the early 1960s, were indifferent or hostile, be it the Bolivian llama herder, the Chilean miner, the Argentine sheepman, or the southwest Atlantic fisherman. As mentioned, there were no strong conservation organizations, no working wildlife reserves on the Patagonian coast or in the puna, and no enforced bag limits for hunters. (There are still no bag limits in Argentina and Bolivia.) Nevertheless, a few national parks had been created. The notion of nature protection was quiescent, yet, many of us hoped, ready to be awakened by the spur of information. Wildlife conservation is the most data dependent of applied sciences, and nature enthusiasts trying to make themselves heard had little knowledge at the time. The surest way to get it, we realized, was to study the animals themselves.

Most of the Southern Cone's big mammals and birds, the larger land and seascape species, are linked by their exploitation by humans for their skins, flesh, eggs, or feathers. Many, often the most ecologically important, reproduce only in colonies, and all of them live in extravagant, often harsh habitats. The mountain flamingos dwell high in the Andes, especially on the wide plain between the east and west ranges of the Andean cordillera called the altiplano. The altiplano is over 3,400 meters (11,000 feet) high, cold, dry, and surrounded by peaks as high as 6,000 meters (19,700 feet). It is characterized by high winds and violent temperature changes—brutal. It extends from southern Peru and Bolivia south along the Chile-Argentina border to the western mountains of Argentina's La Rioja province. There the land descends into the severity of the steppes of Patagonia. In fact, the scant altiplano vegetation is predominantly that of the Patagonian steppes, which drop from the Andes in a series of plateaus all the way to the Atlantic shore. The steppes extend from the Rio Negro (or Rio Colorado) south to northern Tierra del Fuego, covering over 750,000 square kilometers (290,000 square miles).[1] Everywhere, they are cold, mostly dry, and home to guanacos, rheas, and pumas (*Puma concolor*). And where they meet the Atlantic, they do so at an equally distinctive coast.

Remarkably, wonderfully, most of Patagonia's big sea mammals and birds can once again be seen along that coast today—their calls reverberating from high cliffs and rocky shores—even if not in their original hundreds of thousands. Few shores in the world possess so spectacular an assemblage of seaside wildlife as still survives in Patagonia. The challenge is to preserve that diversity and allow it to flourish in the years ahead. Nowhere else is there such a great water's-edge spectacle so accessible for enjoyment and study: a uniquely evocative natural heritage that sits on the edge of a 2,000,000-square-kilometer (772,000-square-mile) ocean region of biological productivity, now called, in one of those excesses of scientific eloquence, the Patagonian Large Marine Ecosystem. This huge subdivision of chilled seawater is bordered at its southeast corner by the 420-island Falkland/Malvinas Archipelago, a rich meeting of the wildlife communities of the Patagonian coast and the Antarctic.

Thus, there are mountains, steppes, seacoast, and a Patagonian sea to consider. The story of their wildlife and its relationships with people falls readily into three more or less animal-defined geographic divisions: steppe and altiplano; coast; and the southwest Atlantic's world of sea and sky.

In conservation biology, the task is to learn enough to be able to predict the probable consequences of various actions. This is not only an informational challenge but also an inspirational one—to inspire long-term public values, to change minds, and to overcome indifference. Although the wildlife of the Southern Cone is sparse compared with that of tropical forests and coral reefs, many of the ways humanity uses it have the same lethal result: overhunting, habitat destruction and fragmentation, and the introduction of invasive species, for example. Most entrepreneurs new to the Southern Cone find themselves sloshing about in an ill-considered bilge of short-term development that clashes with both the long-term economic viability of the region and the new conservation consciousness.

When attempting to explore the nature of complex subjects, the media often tend to proceed indirectly, by asking people their opinions. The result, of course, is information about opinions on a subject but not necessarily about the subject itself. Biologists go to their subjects—the guanacos, penguins, maras, sea lions, and toothfish—and try to get the facts firsthand. The experience of wildlife in the Southern Cone has been fascinating, they've learned, and it has much to tell us—about that wildlife and about ourselves. The story that emerges is a surprisingly intriguing natural and human history, one that, if it does not give us the *right* to be hopeful, nevertheless suggests some good reasons to be.

# *Steppe and Altiplano*

*The Patagonian steppelands descend all the way from the edge of the Andes to the Atlantic Ocean and cover nearly 300,000 square miles of desert plains yet remain little populated. They are home to guanacos, rheas, pumas, and mara—all now marginalized by sheep. Above the steppes is a breathtakingly beautiful plain, the altiplano, where a string of spectacular shallow lagoons cradled between the east and west ranges of the Andes is illuminated by Earth's rarest and most exquisitely beautiful flamingos.*

*Your sheep that were wont to be so meek and tame,*
*and so small eaters, now, as I hear say, be become so*
*great devourers, and so wild that they eat up and swallow*
*down the very men themselves.*
Thomas More

---

# I
# *The Camel and the Sheep*

Pulling into a Patagonian YPF station to fill our car's tank with nafta, Kix and I witnessed a delightful rough and tumble "king of the mountain" contest between a brattish boy, probably six or seven years old, and an equally brattish guanaco chulengo (baby) on a pile of sand. The boy would charge to the top of the pile, face the little guanaco, and throw his chest out. The chulengo would race up the mound and butt chests with the boy, often winning the sandpile kingdom, and the two would tumble down the pile together and then do it all again.

Orphan chulengos are a frequent result of guanaco hunts and are often adopted by rangers at their lonely outposts in Patagonia's far-flung reserves—with varying consequences. A tame guanaco, like the tame gorilla, makes a fascinating pet when it is small and a downright intimidating one when it grows up.

Guanacos are the South American version of the camel, big cinnamon brown animals weighing 200 to 265 pounds and standing about four feet tall at the shoulder, the largest native land animals in Patagonia and the most conspicuous since the ancients: the mylodons, megatheriums, and prehistoric horses and camels. Along with the rhea, they, more than any other species, symbolize the steppes—and their status tells us of its status. Paleontologist George Gaylord Simpson found them "stupid" and "awkward," but a stronger case could be made for Charles Darwin's characterization: "elegant." Today, the guanaco is the Patagonian hunter's target of choice—in U.S. terms, his white-tailed deer—but with no real regulation of the hunt.

Much of the Patagonian steppeland is covered by vast bushy areas of aromatic mata negra, of quilimbay, algarrobillo, wild thyme, and jume, alternating with great bare plains where the vegetation is just high enough to put seeds in your socks. Guanacos appear as big cinnamon silhouettes amid the gray-greens and pale yellows of this desert. Like Pleistocene signposts, they give the landscape scale and history and enhance both its loneliness and its beauty. On a clear evening, they seem larger than life, posing picturesquely atop cliffs and outcroppings, strutting and posturing with a strange grace that makes them impossible to ignore.

East of WCS's camp on Peninsula Valdés, in an area of high yellow sand dunes littered with shiny chips from Tehuelche arrowhead knapping, lives a guanaco troop that sometimes allows me to approach within a hundred yards or so. Their territory is filled with hollows dug out for dust baths and with their curious latrines, piles as much as nine feet in diameter and a foot deep of fecal pellets like big coffee beans. At first, I thought the latrines served as signposts, for many mammals mark their territories with urine or feces. That enables a male to advertise the borders of his "property" and to convey messages with fecal pheromones. But the guanaco latrines near camp are communal. Argentine ecologist Ricardo Baldi thinks that their concentration may be a way to avoid spreading disease.

An orphan female chulengo that adopted Graham and Pat, named Efe, demonstrated the strength of her inherited toilet training by heading for their recently constructed indoor bathroom:

> The first time she did this, a day or two after her arrival, we chuckled at the coincidence. But Efe knew that was the place where she could

> relieve herself, and from then on never went anywhere else, if she could help it. Sometimes she would charge in from outside—in an obvious hurry—and rush to the bathroom.
>
> This charming habit did have its problems. She never actually used the toilet; in fact, usually she felt it was enough to stick her head in and then relieve herself. Of course, the other end of a guanaco, even of a baby guanaco, is quite far from its head. . . .[2]

At the Cabo dos Bahias Reserve along the south coast of Chubut, orphaned Chuli was no less determined and more of a problem. When we recently unloaded our gear into some windowless storage containers cum bedrooms where we planned to spend the night there, we discovered that residence required the approval of this large female guanaco. It quickly developed that acceptance would be unlikely with anything less convincing than a spear or an ax handle. Chuli had been hand-raised by Jorge Cross, the gentle ranger, who now wished he hadn't. She had become 250 pounds of territoriality. A few days before our arrival, and despite Jorge's warning, a large, confident German tourist had approached Chuli to take a closeup portrait and was knocked flat for his pains. Fortunately, he took the experience with good humor, and his camera was not broken. The incident was so worrisome to Jorge, however, that he arranged for Chuli to be blindfolded and driven by truck more than twenty kilometers away. But Chuli is no dummy, no matter what George Gaylord Simpson thought. "She soon identified a tourist bus and followed it, galloping all the way back to the reserve. She arrived dusty, quite put out, and in a fearsome temper," reported Jorge.

Guanacos near the Peninsula Valdés Research Station, somewhat protected from hunting in the immediate surrounds, are curious about us. I have traded whinnies with herd machos for minutes at a time and followed herds through the desert, trying to see what they were eating and doing. I have watched chulengos chase each other, females butt females, and some protracted chases and fights between males. When threats prove insufficient, males rear up on their hind legs, spitting, biting, and pummeling each other with their front hooves—and the victor follows up by chasing and biting his vanquished opponent. But I saw much I did not understand—for example, a sitting female and her chulengo raising their chins toward each other for several long moments in a strangely formal way. Were they vocalizing below my hearing range as elephants do?

Bill Franklin, formerly an ecologist at Iowa State University, has spent years trying to make sense out of guanaco "language" and has uncovered a rich vocabulary.[3] Guanaco behaviors and vocalizations include everything from humming to spitting, as well as submissive crouching and other special body posturing. There is ear and tail signaling, chest ramming, staccato whinnying, dung-pile marking, and the male's melodious mating song, called orgling, during the guanaco's hour-long copulatory bouts (accomplished, as for other camels, while sitting).

Male guanacos, like sea lions and horses, are polygynous; they have several mates at the same time. The main social group is a family composed of an adult male, a number of females, and their young of the year. Although females born in a harem may remain within it, the leading male of the troop usually forces out the young males during their first year, whereupon they gather in all-male herds (there are all-female herds, as well). As usual, females of the species live a comparatively long time. Two female guanacos born at the Cleveland Zoo lived to be twenty-eight years old (M. Jones, pers. comm.). The females are typically as large as or even larger than the males that herd them and defend their territories from other troops and males; it can be hard to tell male and female guanacos apart except by their behavior.

Despite their lack of obvious sexual dimorphism, most male guanacos are highly territorial all year long; they constantly work to defend their troop's food supplies. This gives them access to females for breeding and ensures the health of their families. Territorialism turns out to be a crucial difference between domestic sheep and guanacos and what they do to the land. Guanacos spread out, lightening their feeding pressure on vegetation. Nevertheless, I have seen herds of as many as 140 in the austral winter, and historically, they gathered in herds of thousands. This strongly suggests that they must once have been at least partially migratory before the days of sheep fences.

# ACT III

Not long ago, very large numbers of guanacos ranged throughout the southern half of South America from the Andean foothills of Peru, Chile, and Argentina through Patagonia's desert grasslands, and even into occasional forest. There may once have been forty million or more guanacos in South America. Today, only scattered fragments of that population remain, and the vast majority are on the steppes in Patagonia. With a number of Patagonian colleagues, Graham and I have been trying to figure out how more of the original population, that grand sight, might be restored.

Patagonia's Tehuelches and the Puelche Indians of the pampas north of Patagonia built a culture based on guanacos and rheas that was comparable, in its dependence on meat and skins, to the bison–pronghorn antelope culture of North America's plains Indians, which existed at the same time. Describing a Tehuelche encampment of about 1900, Hesketh Prichard, an English explorer, wrote:

> The most remarkable feature of it was the presence, in one form or another, of the guanaco. Some of his flesh was cooking on a fire outside the tents, the *toldos* themselves were composed of his pelts, the ponchos which some of the women were weaving were made from his wool, the boots were formed of his neck-skin, some of the horse gear of his hide, the man's *capas* [cloak] of his skin, while dogs, men, and women alike were fattened upon the food he provided.[4]

There were so many guanacos and rheas on the steppes that the latter-day Tehuelche Indians captured them from horseback with simple encircling tactics, as described by George Musters: "The ostriches and herds of guanacos run from the advancing party, but are checked by the pointsmen, and when the circle is well closed in are attacked with the bolas [boleadoras], two men frequently chasing the same animal, from different sides. . . ."[5]

Among the world's big, hoofed creatures, it may be that only the American bison (fifty or sixty million) and the pronghorn antelope (forty or fifty million) have exceeded wild guanaco numbers in historical times. African wildebeests have never come close, nor have caribou or saiga antelopes, at least not since anyone has been counting. What is the evidence that there could have been forty million, maybe as many as fifty million guanacos, the vast majority in Argentina, when Europeans settled in South America? Darwin reported seeing very large numbers in Patagonia in 1833.[6] Then there was Musters, who saw herds of three thousand to four thousand in the late 1860s when he was living and traveling with the Tehuelches in the now barren Santa Cruz province.[7] In 1902, the peripatetic Hesketh Prichard wrote that "literally thousands appeared at the summit of surrounding ridges" during his treks through Santa Cruz.[8] But how do we get to estimates of millions?

Starting from the other end, that of calculating food resources and possibilities, guanaco expert Kenneth Raedeke, who developed these large population estimates, observed, "The aboriginal range, though [now] cultivated to a large extent, still supports over 45 million sheep and 25 million cattle."[9] (He was including Chile and much land north of Patagonia, I presume, for I cannot confirm the 45 million sheep.) Thus, the land was clearly once capable of supporting vast numbers of guanacos. Also, the export figures reported in Act II—over 500,000 skins shipped out in the late 1970s alone—indicate that there must have been a great many guanacos.

Deer, antelope, and bison, especially bison, are a part of U.S. history, with a treasured status guanacos have never achieved in Argentina. No longer considered human ancestors, as they were by the Tehuelches, neither are guanacos helpful to sheepherders, hoteliers, or industrialists. While some Argentines might like to live "where the skies are not cloudy all day," few want a home where the guanaco roam and rhea and mara can play, nor are there big steppeland parks where they *could* roam and play. In contrast, any western U.S. park that *can* have bison does have bison—and advertises it proudly. The story of the bison's decline and rescue from extinction has also played a seminal and important emotional role in the development of U.S. conservation ethics.

The guanaco is very different from the bison, of course, and its presettlement hunters are not as admired as North America's Plains Indians. Tehuelches are remembered mostly from ugly turn-of-the-century photos in which they appear surly and degraded in the darkening days of their final disappearance, not as the handsome people that Pigafetta, Drake, and Darwin described.[10] Only now are they beginning to be seen as romantic figures of Argentina's heritage.

The soft, fine skins of very young guanacos are still in demand for bedspreads in Buenos Aires and other large Argentine cities, and they can sell for several thousand dollars. I have repeatedly come upon the carcasses of chulengos killed for that trade, stripped and denuded—pitiful remnants of such beautiful creatures. Graham described his feeling when he stumbled on a chulengo body near the camp:

> Something about it struck me as odd. Then I realized what it was: The fur was gone; only telltale little tufts remained around the top of each hoof. The animal had been skinned. The loneliness of the desert poured in and I became angry. A few yards farther on there was another carcass, and then another. I was witnessing a tragedy that takes place each November, during the season chulengos are born.
>
> The hunt is crude. One or more farmhands set out into the desert on horseback with their dogs. When they encounter a herd of guanacos they chase them until the young ones tire and stop to rest amidst the vegetation. A rider dismounts, corners one of the little creatures, holds it firmly, and slits its throat with a knife. Skinning is accomplished quickly and expertly, the hides are thrown over the back of the saddle and the rider moves on.[11]

How many guanacos are left? Studies compiled by Jorge Cajal and Jorge Amaya in 1985 came up with an estimate of 550,000 worldwide,[12] of which 95 percent are believed to be in Argentina. By way of comparison, there are now upward of 30 million white-tailed deer in the United States, and hunters legally kill almost 400,000 each year in New York state alone. In New York and similar areas, the elimination of major predators and the replacement of primeval forests with farmlands and second-growth woodlots improved conditions for deer enormously. In Patagonia, by contrast, overgrazing by sheep has reduced forage for both sheep and guanacos, creating a disaster almost beyond recovery in many areas, according to Argentina's agricultural institute. Nevertheless, 550,000 guanacos is a sizable number, about the same as the number of U.S. white-tailed deer in 1900. The restoration of big herds can be accomplished — but in Argentina, where almost all land is privately owned, only if the sheep ranchers will tolerate them.

Essentially free of domestic animals in 1885 (except for horses and dogs), by 1910 the dry, windy Patagonian steppe was supporting 12 million sheep. By 1952, Patagonia's sheep population had peaked at 22 million. It has since declined by about 40 percent, to approximately 13 million. I doubt that a habitat challenge comparable to the one wrought by those relentless woolly grazers had swept over Patagonia since it became arid, its plants dry and brittle and slow to regrow. There were millions of guanacos before the sheep arrived, yet the rangeland was ample and self-renewing then. Now, over large parts of the steppes, vegetation has all but disappeared. In less than seventy-five years, the plant community was pawed and nibbled to dust and rocks by sheep, something millions of guanacos had not done in millennia.

Sheep were first introduced to Patagonia in 1876. Some three hundred were brought in from the Falkland/Malvinas Islands, ending up in Santa Cruz.[13] More came from Chile and from Buenos Aires and La Pampa provinces, where they had been displaced by cattle and crops. Sheep ranching thrived in the arid north with Merino sheep, which were bred primarily for wool, and in the south with Corriedale, sheep raised for both wool and meat, which arrived in 1905 from New Zealand. After refrigerated ships became available in 1894 for meat transport, both sheep and cattle farming boomed, and the guanaco, seen as a competitor as well as a hunting target, began a precipitous decline.[14]

In 1889, the province of Chubut had 20,000 sheep. Just seven years later, it had more than 100,000, and, by 1911, it was home to 5,000,000 sheep![15] Today, there are about 4,000,000 sheep in that province, more than in any other Argentine state. Each sheep needs to eat about 3 percent of its body weight in greenstuff each day. Patagonian sheep average about forty kilograms (eighty-eight pounds). So the Chubut steppe must produce between ten and eleven million pounds of sheep fodder every day, about two million tons each year. No wonder it is almost impossible to walk anywhere in the backcountry of Chubut without trodding on the mummified pellets of old sheep dung. Unfortunately, the soil is too dry and thin to support the communities of invertebrates and other soil organisms necessary to incorporate the dung within a humus, to provide for its recycling and make its nutrients available to plants again. It simply becomes part of Patagonia's infamous polvo, or dust. Thus, there is reason to suspect that a fair portion of the Patagonian dust blowing in your face is powdered sheep shit—and, all things considered, there is a certain justice in that.

Sheep are everywhere, but you can drive great distances across Patagonia and see a guanaco only occasionally. Claudio Campagna and Ricardo Baldi have tried to get an aerial perspective on the situation on Peninsula Valdés, which is officially a "reserve." Here is an excerpt from Claudio's description of a 1995 survey:

> The Cessna 182 left Puerto Madryn at dawn. Three researchers shared the claustrophobic cockpit with Pablo Pascual, a young pilot trained in the spirit of Saint-Exupéry. . . . The workplan was fatiguing: fly fourteen transects in about eight hours with only one stop for rest and refueling.
>
> Flying at 300 feet proved to be too low for comfort, but we had our reward, a spectacular view of the endless plains. The pilot even managed to spot Darwin's rheas. (We thanked him very much but begged him to concentrate on the airplane's instruments.) Most of our time was spent recording sheep herds: ten animals here, forty-five animals there, more than 150 sheep ten minutes after departure. The first sight of guanacos came after seventy miles of sheep counting: four adults and one chulengo.
>
> The 1,500 square mile Peninsula Valdés is a wildlife reserve about the size of Yosemite National Park in the U.S. We found that 2,200 guanacos shared the reserve with 80,000 sheep, about one guanaco to every 54 sheep. Extending our survey areas beyond the Peninsula . . . to about 10,000 square miles (an area about three times the size of Yellowstone

> National Park) we got more promising results, about 43,600 guanacos and 240,000 sheep, one guanaco to 5.5 sheep. Guanacos were always confined to the driest, least productive lands in the study area and were most abundant where sheep were least abundant.[16]

Claudio goes on to observe that in a recent ten-year period, farmers were given permits to kill 38,000 guanacos in an area with a population of 12,400 animals. "Entire herds had been eliminated by a know-nothing administration that ignored population parameters, management principles and the permanent range destruction caused by sheep."[17]

During the First World War, the price of wool tripled and meat was in great demand, so even the most marginal Patagonian lands were used for sheep. Wherever sheep were concentrated, guanacos declined or disappeared, along with an entire community of birds, mammals, reptiles, and invertebrates dependent on the original vegetation and each other. And now, since the 1980s, the sheepherders themselves are disappearing. They put themselves out of business by keeping so many sheep that they have destroyed the vegetation on which their business depends. To view it unsympathetically, their behavior replicates the short-sightedness that has characterized so many commercial fishermen, seal hunters, plume traders, whalers, and loggers around the world. Although we don't usually think of farmers in that context, it is time we do—even if it is, as one Argentine official indignantly (and eloquently) claimed, "ideological apostasy."

Continuous year-round grazing, in which sheep roam the range largely unherded, continues to be the Patagonian management, or lack of management, of choice. There are said to be 13,146,000 sheep in Patagonia currently, and nearly 60 percent of the rangeland has been "moderately to severely degraded."[18] In fact, we humans have altered much of the Patagonian "pasture" as severely, if not as intentionally, as if we had selectively bred its vegetation for sheep-resistant varieties and a plantless desert. This sad situation has been played out around the world, the protagonists changing but the plot shamefully the same—although not quite. Patagonia's sheep owners are not desperately poor people long caught in a trap of overpopulation and illiteracy. Different choices could be made.

Recently, Argentine economists calculated that a rancher needs at least 6,000 sheep to sustain an economically viable farm, which suggests a minimum landholding of 20,000 hectares (nearly 50,000 acres) in dry Patagonia. Less than 5 percent of Patagonian landholdings exceed 20,000 hectares, however, so most of the rest try to make ends meet by overstocking the land.[19] And in 2003, the wool and mutton markets suddenly revived—promising more overgrazing.

Wool prices, which had hovered at or below $2 a kilogram for almost ten years, jumped to as much as $9. Prices for mutton more than doubled. Why did this happen? The 2002 devaluation of the peso made Argentina's wool exceptionally cheap, and drought losses made Australia unable to dictate world prices. At the same time, the price of oil, and of oil-based fibers, increased, and mad cow disease in Europe made mutton more attractive to that market.[20] Foreign firms such as Benetton Group profited most. Luciano and Carlo Benetton bought distressed sheep ranches at rock bottom prices and, with two million acres, are now the largest landowners in Patagonia. For wild animals, this could be bad news, for the intensity of Benetton management leaves little habitat for them.

Nearly one-sixth of Argentine Patagonia now belongs to 350 foreign owners. No one expects the Argentine wool business to regain its salad days, but it has received an unexpected reprieve. The few sheepmen who have grass and very big holdings will be okay until the next time the market declines significantly, or the amount of rainfall changes appreciably, or the winters worsen. But many herders recognize that there may never be as many sheep again and that they are headed for serious economic trouble. So some are looking into ranching guanacos, mainly for their wool. Considering the amount of investment, monitoring, and understanding necessary to make a success of guanaco rearing, one must admire their courage. "Guanaco Venture Protects Local Ecosystems," claims an Argentine advertisement for a business in the World Resource Institute's Sustainable Enterprise Program. But a study by wool experts in Italy (where 50 percent of the world's wool products are manufactured) points out that commercial success can be expected only through long-term selective breeding and intense management, and then only if a highly capricious specialty market continues to attach high value to guanaco wool.[21]

Guanaco territorial behavior, so tightly tied to supportable population numbers and resources, is certainly better adapted to protecting the range than that of sheep—and guanacos have broad, soft feet, easier on the land than sheep hooves. The social system of sheep is based on a single dominant leader and predisposes them not only to selective breeding and a home range, but also to massive crowding—they do not defend a territory. If guanacos could be convinced to behave the same way, it would make farming and selectively breeding them much easier, but the benefits of their spaced-out territorial system for preserving habitat, and a genetic diversity responsive to their wild habitat, would then be lost. They, too, would degrade the land. Ricardo Baldi observes that the vast majority of the plants they eat are the same as those sheep eat.[22]

While it is a dubious hope that more pounds of guanaco (and more valuable guanaco wool) than of sheep can be produced on badly degraded rangeland, guanaco wool is much more valuable in the current market. Ricardo Irianni, a Chubut businessman, says that although one can shear only about half a kilo of wool from a guanaco at one time, compared with three kilos or more from a sheep, guanaco wool is selling for $100 per kilo. Guanaco boosters claim that the wool is very fine, the hairs 16 $\mu$m to 18 $\mu$m in diameter, but the Italian experts say that such fine quality is rare and insist that it can be reliably produced only by selective breeding, contradicting the guanaco proponents' claim that guanaco wool approaches the quality of vicuña's wool among cameloids, 10–12 $\mu$m (cashmere is 13 $\mu$m). Nevertheless, right now, two guanacos could produce as much income as five sheep.

It is worth remembering that today's sheep are the result of hundreds of years of selective breeding. (It is claimed that Merinos were developed during the reign of Claudius, AD 14–37.) They are easily fenced, produce much more wool than guanacos, breed fast (gestation is about 5 months), and are docile and easy to handle. Guanacos have a gestation period of 11 months (345–360 days), are hard to confine, and can run at speeds of up to 37 mph—Graham and I have clocked them. Even tame ones can be difficult to handle. I will not soon forget Chuli or the experience of giving a zoo guanaco an enema—nor will any of the six battered and spit-upon fellows who helped me.

Bill Franklin, one of those arguing for ranching guanacos, quotes the famed ecologist Norman Myers: "The more we can demonstrate the economic value of wild creatures, the more we shall add a much-needed weapon to our arsenal of conservation arguments." Well, yes, some species, sometimes, some ways. But let us not forget the qualifier, "wild creatures." From the standpoint of the *wild* guanaco's survival, the value of farming depends on how it is done. The kinds of guanaco management that will be most rewarding to the farmer will almost certainly require domestication over time: selective breeding for docile, rapidly reproducing animals with the best wool. This means that, in a few generations, we would produce glorified llamas, not wild guanacos—just as we have domestic cows, not wild aurochs; poodles, not wolves; chickens, not junglefowl.

Perhaps the key question for both ecologist and farmer is, Can guanacos survive as a wild species and be ranched in marketable and cost-effective numbers on land badly degraded by sheep? From the perspective of the wild guanaco, the best species survival approach would be open-range ranching, big spaces able to accommodate lots of free-ranging guanacos unfettered by domestic selection, more or less the way Ted Turner is handling North American bison. In fact, Turner, a Patagonia enthusiast, is very interested in the guanaco question and expects to try some experiments on his Argentine properties. But guanacos are different from bison, and few Argentine sheepherders are likely to be able to follow his example. So one could breed guanacos for wool if it is economically sound but should not claim that it is helping the ecosystem or the wild guanaco. Unless the breeder is sustaining and sensibly cropping wild populations, it probably isn't.

The Cabo dos Bahias Reserve, Chuli's home, near the village of Camarones, is an exceptionally beautiful place to watch guanacos, as many as five hundred until recently. They were all over, often standing on the park's spectacular red granite rock formations silhouetted against the sea. With less than three thousand acres, however, the reserve is much too small to sustain such a population, even though some jump fences to nearby estancias. When a harsh, dry summer was followed by a very cold winter in July and August of 2000, the Cabo dos Bahias guanacos starved; four hundred were lost, bringing Jorge, the ranger, to tears with the memory. He took us to view the bodies. They lay as they had fallen, huddled under bushes in family groups, females and their chulengos together. Examination by Jorge, WCS veterinarian Marcela Uhart, and Ricardo Baldi found few males; it was families of females and young that had stayed and died, the chulengos attempting to suckle on their mothers even after they were dead, and then dying beside them.

Such starvation is not new to guanaco populations. Darwin found what he thought might be guanaco cemeteries:

> The guanacos appear to have favorite spots for lying down to die. On the banks of the St. Cruz, in certain circumscribed spaces, which were generally bushy and all near the river, the ground was actually white with bones. On one such spot, I counted between ten and twenty heads.[23]

Starvation may be an increasing problem now that most of the guanaco's range has been fenced, restricting movement, and degraded by sheep. But to hear some local ranchers tell it, you might think that a guanaco's prime function is to be the major source of sheep diseases. Fortunately, well before the guanaco famine, I had asked WCS's field veterinarians, Billy Karesh, Bob Cook, and Marcela Uhart, to test some wild guanacos and domestic sheep. They did so at Cabo dos Bahias.[24] The team found that the guanacos were negative in all of a wide variety of serologic tests, whereas the sheep were positive in four tests. "Chances are," Billy concluded, "that sheep are a source of disease for guanacos, not the other way around." Historically, diseases from domestic sheep have extirpated populations of wild bighorn sheep, those from poultry have infected wildfowl, a domestic cattle infection (rinderpest) has devastated African antelope, and farmed fishes are now infecting wild salmon. The list of such infections is growing worldwide.[25]

From annihilating guanacos to ranching them, from accusing

guanacos of infecting sheep to accepting that it is the reverse, human-wildlife relationships on the steppes, even as Patagonians inch into Act III, would bemuse the Tehuelches: "Well, of course! Guanacos were the backbone of our economy, along with rheas." Any long-term ranching of the overgrazed steppes will depend on restoring and sustaining their vegetation, and it does seem probable that guanacos will play a role. Along with that of the more sensitive Darwin's rhea, their presence provides a clear indication of land's well-being and productivity, even now.

***bå•rom′e•têr****, n. [Gr.* baros, *weight, and* metron, *measure]*
*1. an instrument for measuring atmospheric pressure and thus forecasting the weather or finding height above sea level.*
Webster's New Twentieth Century Dictionary

---

# 2
# *Mr. Darwin's Ostrich and the Green-Egged Martineta*

Not long ago, the Darwin's rhea numbered millions and millions on the Patagonian steppes. More than any other creature I can think of, this big bird, by its number and distribution, indicates the health of the steppelands environment—a steppelands barometer, or miner's canary. It has proved so sensitive to overhunting, excessive disturbance, invasive species, and habitat destruction that the state of its population is a litmus test of the health of many steppeland animal communities. It wanders widely through the ranges of many other steppe creatures, making it what conservationists call a landscape species: Meeting its living requirements is critical to the well-being of a broad spectrum of other species in its region. Darwin's rhea is Patagonian, and its largest numbers have always been on the dry steppes, but a mountain form of the bird, the suri, or Andean rhea, exists in the high, dry regions of northwestern Argentina and adjacent parts of Bolivia and Peru, though it is now rare. Wherever found, the bird is dignified and bizarre. Here's how Darwin described the discovery of his namesake:

> When at the Rio Negro in Northern Patagonia, I repeatedly heard the gauchos talking of a very rare bird which they called the Avestruz Petise [smaller ostrich]. . . . When at Port Desire [Puerto Deseado], in Patagonia (lat. 48°), Mr. Martens shot an ostrich, and I looked at it, forgetting at the moment, in the most unaccountable manner, the whole subject of the Petises, and thought it was a not full-grown bird of the common sort. It was cooked and eaten before my memory returned. Fortunately, the head, neck, legs, wings, many of the larger feathers, and a large part of the skin, had been preserved; a very nearly perfect specimen has been put together, and is now exhibited in the Museum of the (London) Zoological Society. Mr. Gould, in describing this new species, has done me the honour of calling it after my name.[26]

Such is the nature of zoological discovery. But naming honors can be transient. Scientifically, the bird was originally named *Struthio darwini*. When it became clear that rheas are not ostriches and Darwin's rheas are not the same as the greater rheas (*Rhea americana*) of the pampas in the north, its genus was changed to *Pterocnemia*. Even the specific name was dropped; Darwin lost out, and the bird became *Pterocnemia pennata*. There the matter rests, at least insofar as some taxonomists are concerned, but daily usage is more lively. Various sobersides reflected on the bird's size (about fifty-five pounds, smaller than the "greater" rhea found in the north) and changed the name from Darwin's rhea to "lesser rhea," reminiscent of greater and lesser flamingos, greater and lesser kudus, and so on. Such comparative names may work well for a museum taxonomist looking at a tray of skins, but they can be a bit annoying to the biologist who spies just one live bird in the field—"lesser" than what? Nowadays in Patagonia, the species is often called choique, ñandú petizo (little rhea), and sometimes Darwin's rhea. Whatever it's called, it's a wonderfully weird creature, and its original abundance, food requirements, and seed dispersals—through its rather casually digested droppings—must once have been a major ecological force on the steppes.

Darwin's rhea swims and can run nearly as fast as a guanaco—on only two legs, and with virtually no defenses—from the Atlantic seashore to the edge of the Andean altiplano. In comparing rhea and guanaco, I am reminded of the observation that Ginger Rogers did everything that Fred Astaire did "backwards and in high heels." For me, the choique's unexpected silhouettes quietly grazing in small flocks on the lonely steppe arouses thoughts of timelessness, of a primordial being little changed.

The Tehuelches used the birds in every imaginable way. The feathers were used for decoration and later, as we have seen, as an article of commerce (mostly for feather dusters), fetching about a dollar a pound in Buenos Aires in Musters's day. Marrow from the leg bones was used for making a sort of pomade; sinews from the bird's legs were used in making boleadoras; and the neck was used to make pouches. Rheas and their eggs also suffered the distinction of being a favorite food of the Tehuelches. Cooking was simple: A small hole was made at one end of an egg, a stirring twig inserted, the egg propped up at the edge of a fire, and an omelet for four was soon ready. Adult birds were plucked and split open, hot stones placed in the body cavity, and the entire bird placed in the fire. Unfortunately, the Indian taste for rhea is one that gauchos and sheepherders maintain to this day, which contributes to its precarious status in most places.

The vast majority of what we know about these birds has been learned from studies of the related greater rhea of the pampas, and some of the most intimate details were worked out by my colleague Donald Bruning.[27] Don was curator of birds at the Bronx Zoo for many years and studied greater rheas there and in Argentina. He grew to know the zoo birds so well that he could predict, almost to the minute, when a female was going to lay one of her beautiful twenty-four-ounce, ivory-colored eggs (not pale green like the fresh eggs of the Darwin's rhea, which fade to a dirty white). Don would look calmly at his watch, excuse himself, pick up his camera, and saunter over to the zoo's rhea pastures just in time to record the event. He found that egg laying was also dependable in Argentina, mostly taking place between 11:00 a.m. and 3:00 p.m. Rheas reverse the usual sex roles—the male incubates the eggs and cares for the young, as well as preparing the nest. In a series of experiments, Don found that the birds were discerning about egg color. Incubating males would accept artificial white eggs, yellowish eggs, or pale green eggs, but they would not take eggs of more fanciful colors. Unlike many geese, they would not accept square substitutes, either.

Greater rhea courtship begins with deep booming calls, almost too low to hear, "ñaannn-du," and the male displays to a female by zigzagging around her, extending his wings, shaking his feathers, and booming with a thick, inflated neck. Darwin's rhea courtship is similar, but its calls seem even deeper and more prolonged. Although males fight for dominance, they defend only the area immediately around their nests or around their females before nesting, Don says. A dominant male copulates with several females.

Females approach the nest as a group, and each lays an egg near its edge, one after another, which the male rolls into his nest. They return over several days and lay more, but then they depart for other males with which to mate and lay more eggs—they switch mates. Technically, males are "simultaneously polygynous," while females are "serially polyandrous." Gustavo Fernández and Juan Reboreda, of the University of Buenos Aires, have shown that this breeding system puts such strain upon the males that fewer than 20 percent will attempt to nest during a breeding season.[28]

One result of the rhea's breeding system is that a single nest can contain as many as forty eggs. Incubation usually takes between thirty-six and thirty-seven days, but it can be as short as twenty-nine days or as long as forty-three. Nevertheless, all of a nest's chicks hatch at more or less the same time. The chicks achieve this remarkable synchrony by calling from their eggs when they are ready to hatch, by listening, and by harmonizing their hatching times. Some must hatch with little more than thirty days of incubation, while others must wait for over forty days, until they hear the calls. While this seems impossible, it is essential in the rhea system, because the male leads the hatchlings away from the nest within twenty-four to thirty-six hours of the first chick's hatching.

After hatching, the chicks depend on their father's care for about six months. They cluster about him or close by as he leads them from the nest, feeding continually but always alert to danger. When he stops, they bunch up at his feet, and he may brood the lot, covering them with his wings to keep them warm. Although he leads them from feeding place to feeding place, he never actually feeds them. They must forage for themselves from the first. Nevertheless, he responds immediately to their plaintive whistles when they are lost and attempts to defend them from predators, despite his lack of weapons other than his feet (for kicking or trampling), which are useful against only the most modest of enemies.

Although flightless, the "dads" will try to lure predators away from their broods with an amazingly zealous broken-wing act or distraction display. When faced with a predator too difficult to fight, a father rhea spreads and drags his wings in the most distressed fashion, crouching and pulling himself along the ground as though badly wounded and about to expire on the spot—yet always just out of reach. He thus entices the predator (for instance, me while I was making a movie) away from his chicks with the promise of an easy meal. The sham is maintained until the male has drawn the enemy a fair distance—then he jumps up and runs like the wind, later rounding up his chicks.

Most farmers consider rheas beneficial because of their function as seed dispersers. One comes upon "bathing" spots, small versions of guanaco dust baths; upon mysterious intersections with the tracks of other rheas, tinamous, and foxes; and upon the rhea's remarkably undigested droppings. When the thornlike quilimbay (*Chuquiraga avellanedae*) is in blossom, its branches adorned with hard, sharp yellow flowers, choiques walk slowly among the bushes, browsing and swallowing the formidable blossoms with the persistence of a dinnerless diplomat gulping hors d'oeuvres at a cocktail party. Eventually, they pass these prickly inflorescences, thorns and all, in damp compactions that are scarcely changed. It is unclear what nutrition quilimbay provides and downright painful to think about it. Graham wonders if some seeds need to be passed by a rhea in order to grow.

A recent "official" count of choiques, I was told, claimed a population of 1.7 million, but it was based simply on asking ranchers how many they thought lived on their estancias. In Neuquén province, this census claimed 2 choiques per square kilometer (1 every 124 acres).

Actual counts by trained field biologists found only one-fifth as many, only 0.4 per square kilometer, however, and noted an 86 percent decline in about ten years. Rhea expert Jorge Cajal found the high-mountain race of Darwin's rhea, the suri (*Pterocnemia pennata garleppi*), to be rare, even in the remote San Guillermo Reserve in La Rioja, and warned of its deteriorating status. A high level of hunting and predation has long been the rhea's lot. In 1889, famed ornithologist P. L. Sclater wrote, "They were formerly very abundant along the Rio Negro; unhappily, some years ago their feathers commanded a very high price; Gauchos and Indians found that hunting the Ostrich was their most lucrative employment; consequently these noble birds were slaughtered in such numbers that they have been almost exterminated wherever the country admits of their being chased."[29] What seems to be causing their decline now? In some places, it may be introduced European hares and sheep.

Hares don't eat rhea chicks, but foxes (the coyote-sized culpeo, *Pseudalopex culpaeus*, and the little gray fox, *P. chilla*) do. Those predators have preyed on rhea chicks for millennia, however—so what's new? Besides reducing the rhea's food supply, sheep provided new game for the culpeo, which has meant more culpeos—despite more culpeo hunting and poisoning by sheepmen. And now hares have become available, millions of them, also eating the rhea's food supply. And even if the culpeo runs out of sheep, it has plenty of hare to eat—which further increases the culpeo population. Perhaps more important, the hare's young are a perfect meal size for the abundant gray fox and for the larger hawks, as well. Vulnerable little rhea chicks have to contend not only with human feather and egg hunters (still), but also with all their old enemies in larger numbers than ever before.

For the wildlife of the steppes, home on the range has thus changed. The foods of the herbivores—rheas, guanacos, maras and other rodents—and many small birds are being expropriated by sheep and hares, while the predator community's larder is augmented by the usurping sheep and hares, so there are more predators. The steppe's rhea barometer forecasts bad conditions for wildlife.

Giving the picture yet another twist, rheas are now being farmed for

Mother guanaco with her "chulengo," Torres del Paine, Chile.

Guanacos gather as night falls, Punta Tombo, Chubut, Argentina.

Guanacos walk warily across Payunia Reserve, Mendoza, Argentina.

Darwin's rheas, except for guanacos, are the dominant original species of the steppes.

Darwin's rhea male with newly hatched chicks, Torres del Paine, Chile.

Magellanic penguins threaten Darwin's rhea near their nest, Punta Tombo, Argentina.

Sheep, unlike guanacos, accept dense crowding, Santa Cruz, Argentina.

Gauchos: Argentina's famously expert horsemen, Neuquen, Argentina.

A martineta, Peninsula Valdés, Chubut, Argentina.

Glossy martineta eggs reflect me as I photograph them, Chubut, Argentina.

Chaco tortoise, "Millicent," from Rio Negro Province, Argentina.

Maras, Peninsula Valdés, Chubut, Argentina.

Punta Bermeja's cliffs, home to barranqueros, Rio Negro, Argentina.

Barranqueros at the El Cóndor nesting cliffs, Rio Negro, Argentina.

Subtle differences in size distinguish male and female barranqueros. Here, the male is in the foreground.

Argentine gray fox, Punta Tombo, Chubut, Argentina.

meat and eggs. Darwin's rheas are easily farmed, and special qualities are claimed for its meat (low fat, low cholesterol), so farming has started in Argentina and, for the greater rhea, even in the United States and Canada. One advertisement, "Granja Choique Malal & Loma Blanca" (taken from the Web), describes the operations of a Patagonian ranch specializing in farming Darwin's rheas. It claims to be the world's first commercial Darwin's rhea farm and to have already raised four generations of choiques between 1994 and 2000. The ranch is advised, the advertisement notes, by experts from the National University of Cordoba, and it offers for sale fertile eggs, chicks up to six months old, "meat, hides, grease, feathers, whole egg shells, etc." Choiques are also being farmed in Chile, where captive-bred chicks are implanted with individually coded microchips to ensure against illegal exploitation of the wild population.

At H. C. Macfie's Pampas Poultry farm in Utah, selectively bred white (greater) rheas are offered. Mr. Macfie provides recipes on his Web site but suggests that rheas are "friendly and rather endearing" and can "simply be kept as a pet." Comparing the farming of ostriches to that of rheas, Mr. Macfie comes down convincingly on the side of the latter: "Rheas are easier to get on with because they are smaller than you, whereas ostriches are bigger than you."

Of course, such farming efforts are not "saving the wild rhea," but rhea farmers make no bones about that. They are in the business of rhea farming for its products. Inevitably, the farmer initiates selective breeding for features that make the birds more marketable and easier to farm—and less suitable for the steppes. To produce birds representative of the wild populations, the farmer would have to avoid traits that make them bigger, fatter, more trusting, less aggressive, and so on. To sustain wild rheas, he would have to manage his breeding stock to maintain as much original genetic diversity as possible, as zoos do in trying to maintain the wild species' heredity.

Also being farmed, experimentally, are some of the tinamous—strange partridge-sized birds distantly related to rheas. For example, the producer of the glossy green eggs I found on my first visit to a Patagonian shore is the martineta, or elegant-crested tinamou

(*Eudromia elegans*). It is not endangered. It shares some of the rhea's history and some obscure anatomical traits, as well as a reversal of the usual sex roles. However, its extraordinarily conspicuous shiny green eggs distinguish it from all other Southern Cone birds (although most of the other forty-six species of tinamous lay indiscreetly bright glossy eggs: baby blue, bright pink, red vinaceous, pale lilac). Tinamous range from quail size to chicken size but are entirely different from such gallinaceous birds. Among other differences, the lungs and heart are tiny. A domestic cockerel's heart accounts for up to 12 percent of its total body weight, a tinamou's only 1.6 to 3.1 percent. No other bird has such a proportionately small heart. Nevertheless, our green-egged martineta is abundant. In an hour's drive on Peninsula Valdés, I counted over seventy foraging for seeds and insects along the road and heard the musical whistles of many more.

Although martinetas are polygynous like rheas, females strongly dominate the males. It works like this: Two or three females form a breeding coalition and wander from one male to another. Once they have selected a male, they assert their dominance by chasing and pecking. When the male has built a nest, they simply drop off their eggs for him to incubate (19 to 21 days) and walk off to seduce and henpeck another male—as many as seven or eight during the summer. A single female may lay as many as thirty-five eggs in several nests. One at the Bronx Zoo laid fifty-five eggs in 124 days, each weighing about an ounce and a half, more than five pounds worth altogether—about three times the weight of the female. Besides its fecundity, the bird's success rests in cryptic coloration. Its black and white and gray feathers and pattern of bars and stripes make it nearly impossible to see in dry thornbush, and it relies on that advantage. When martinetas are visible, they look slightly goofy; disproportionately short legs support a plump, round figure, long, thin neck, and a head surmounted by a tall, thin crest, usually slightly awry.

Again like rheas, martineta chicks hatch all at once and leave the nest when only a day old, following their father in a constant search for food. They stay with him for about a month; when food is abundant, a male can rear as many as three broods in a season. If a brood is surprised by a predator, the chicks scatter and freeze. When startled, a father with his newly hatched chicks may go into a desperate broken-wing act, similar to the rhea's. But tinamous can fly, so, when a predator has been drawn away from the chicks, the father flies back to them. The chicks freeze with such confidence in their protective coloration that I have had to search on hands and knees to find one I knew was there—which was eventually betrayed by reflections in its bright, lovely eye.

The ancestry of tinamous and rheas is among the oldest of all the steppeland's inhabitants. Rheas appear in the fossil record at the beginning of the Tertiary, and both are probably derived from a lineage that extends back at least 100 million years. Our Earth has circled the sun 40 million times since their kind first walked upon it. Along with the guanaco and the mara, they are more evocative of the wild Patagonian steppes than any other creatures. They were very special to the Tehuelches, and I wonder if they could become special to the new Patagonians. Is it too romantic to attempt to restore a few major wild populations, symbols of Argentine Patagonia's recovery from land abuse and of a new concern with its unique inheritance of steppeland wilderness? Could there once again be millions of choiques as well as guanacos?

Counting Mr. Darwin's "ostrichs" is a convenient way to measure the persistence of wildness on the steppes and probably provides a more sensitive gauge of the parity of its predators than counting guanacos—although the guanaco's appetite makes it a better indicator of the recovery of steppeland forage. They are both landscape species. If the choique can increase its numbers to some semblance of its past abundance, we will have convincing evidence that what it eats has revived and what eats it is again in some sort of equilibrium. Then if a giant, uniquely Patagonian mouse called the mara can recover, even the Tehuelches would feel at home.

*In order to enjoy a dog, one doesn't merely train him to be a semihuman. The point of it is to open oneself to the possibility of becoming partly a dog.*
Edward Hoagland

---

# 3
# *A Giant Monogamous Mouse*

If while traveling in the steppes, you spy a sleek, elegant mammal about eighteen inches tall that dashes off through the bush "stotting," jumping high in the air with all four feet, like an African antelope, you are not hallucinating. You have seen a mara (*Dolichotis patagonum*). It is a sort of huge mouse, a giant, long-legged rodent that is faithfully monogamous yet rears its cubs in communes. If you see one in the wild, you really are in Patagonia; it is found almost nowhere else (except for a few in midwestern Argentina). Darwin thought the mara a kind of hare, but Commander Musters considered it a small deer. The confusion is understandable: The mara has the erect ears of a hare but long, slender, deerlike legs, and its feet are small and almost hooflike. Reproductively, no Patagonian animal could differ more from rhea, tinamou, and guanaco than the mara. No animal is more typically Patagonian, and, in the beginning, none caused more confusion about its nature.

Much of what we know about mara behavior is the result of remarkably intimate studies of a captive collection in France.[30] Maras do well in zoos; a Bronx Zoo colony had sixty-seven births in thirteen years.

The only investigation of maras in the wild, however, was conducted on Peninsula Valdés in the early 1980s by Andrew Taber.

Andrew is strong and compact with deep-set gray eyes under a broad forehead surmounted by an Einstein-like mane of prematurely gray hair. He thrives on modest accommodations and major challenges. In a self-deprecating way, he admits that he sometimes looks "slightly mad"—especially in the middle of an enthusiasm. If he were, it might explain how he has been able to live for months on end in the Patagonian bush, or the Paraguayan Chaco, or the Bolivian rain forest.

The son of a U.S. foreign service family, Andrew was born in Taiwan and shuttled back and forth between Japan, Holland, Malaysia, and Vietnam while he was growing up. After surviving the University of California at Santa Cruz, he fled to Alaska to count whales and to the Aleutians to count seabirds; obtained a doctorate at Oxford; and ended up, naturally enough, he claims, in Patagonia—for a while. He has a poorly controlled hankering for adventure and, since joining WCS's staff, has worked in Paraguay and Bolivia and accompanied me on new projects to remote islands in the Falkland/Malvinas. He connected with Patagonia when he decided to study that twenty-pound mouse, the mara.

To study maras in nature, Andrew spent three years sharing one-room huts with a series of tolerant puesteros on Peninsula Valdés ("Two hundred and ninety hours were spent observing maras at den sites, one hundred hours studying them away from den sites, weekly surveys . . . ," etc.). As his hosts took care of their sheep, Andrew figured out the social system, ecology, and love life of maras—and fell in love with and married a lovely Argentine woman (no hours recorded). No doubt his family was further impressed with the pull of Patagonia when his sister, Sara, married a field scientist and moved to Peninsula Valdés, near Graham and Pat at the WCS station. (Sara's husband, Peter Thomas, studied whales, while she, a sociologist, investigated the lonely lives of the isolated wives of the sheep farmers scattered throughout the campo.)[31]

Andrew's study, undertaken for his doctoral dissertation at Oxford, sought to answer two fundamental questions: Why are maras monogamous? Why do they den communally? Though these questions may have little to do with mara conservation at first sight, they are directly related to mara territorialism and fecundity, hence with population size, and so with conservation. A social system that incorporates both monogamy and communal denning is unknown in other mammals. In fact, few mammalian species are monogamous, and among those that are (gibbons, South American night monkeys, and the like), even fewer are as monogamous as the mara. Generally, birds are better at monogamy—and slyer at extramarital liaisons.

Andrew's methodologies were direct observation from blinds and radio tracking. Here, condensed, is one of my favorite scenes from his written observations:

> Dawn broke across the Patagonian thorn-scrub as a female mara cautiously approached a den, followed closely by her mate. At the burrow's entrance, the female made a shrill whistling call, and almost immediately eight pups of various sizes burst out. The youngsters were hungry, not having nursed since the previous night, and all thronged around the female, trying to suckle. Under this onslaught, she jumped and twirled to dislodge the melee of unwelcome mouths, which sought her nipples. She sniffed each carefully, chasing off those that were not her own. Finally, she managed to select her own two offspring and lead them away from the crowd for an hour of intermittent nursing, despite the harassment of other hungry infants. During all this her mate sat vigilantly nearby.[32]

It seems a stretch, but this curious creature is related to the common guinea pig, which was domesticated for food by the Incas as early as 1530 and is used in laboratories and kept as a pet all over the world. But the mara has never been domesticated. More than thirty inches long, it is sizable and strong and can run 40 mph for several hundred meters.

A mara is meal sized for culpeos, or pumas if they can catch one, and it seems strange that its escape behavior features the bouncing stotting jumps of some antelopes, such as African springboks or impalas. It's as though the mara is jumping above the vegetation to keep an eye on its pursuer or, as some have suggested, demonstrating its superb condition to convince the pursuer of the futility of trying to catch it. In any case, its antelope-like behavior provides a remarkable instance of convergent evolution. In the mara, the visual effect is emphasized by the attractive "mini-skirt" of prominent white fur that fringes its bottom and makes some say it looks better from the back.

Despite the mara's escape technique, and the difficulty of getting close enough to take a photograph, they are often unreasonably tame. On one remarkable occasion, when I was crouched in some thornbushes trying to see what a matuasto, a small but formidably strong-jawed lizard, was doing, a mara joined me, sat down, and watched for almost five minutes. Animal observation is not a one-way street. Crouching, getting your eyes, nose, and ears at a smaller animal's normal level, is always edifying, another way of trying to enter a smaller creature's world. You know that you are accepted when you hear an unexpected noise and, simultaneously, both look in the same direction, as the mara and I did.

Gradually, Andrew's notebooks filled with mara natural history. Maras turned out to be selective grazers, seeking only the most nutritious buds, flowers, and, especially, grasses. They feed largely on grassy patches in the little open areas between bushes. A single pair could exhaust its selected food supply in an area of fifty square meters (over half an acre) in an hour, so foraging kept them on the move nearly half their day. That means they have to space themselves out in order to find enough food without competing, one reason for moving about in pairs rather than groups. Insofar as male-female relationships go, the fact that you must put your head down to graze is important. The females, which have to fuel the demands of pregnancy and lactation, graze more than the males, so it is helpful to have a mate with his head up, vigilant.

There is another reason for male maras to be so faithfully monogamous. Female maras have an extremely short period of sexual receptivity, usually only a few hours once a year. If the male wishes to breed, he had better be there at that time. Nevertheless, the male mara's faithfulness and constant attention are impressive. The French investigators with their captive maras noted that the mate of a female that was blind in one eye always stationed himself protectively on that side. Another male helped his partially paralyzed mate to run.[33] Andrew says that Patagonian sheepherders hold up the mara as an example of familial loyalty and claim that if one kills the female mara, one must also kill the male because it will never mate again. It is at least true that if a male loses his partner, it may be a considerable time before he forms a bond with another, despite the availability of eligible females.

A pair usually stays within a few yards of each other, the female setting the pace and determining the activities. Their bond is sustained by the male, a faithful follower. Should the male be separated from the female, he will sound a high whistling note. Bonded pairs also "emit a low grumble" alternately and almost constantly while grazing and moving, Andrew has learned. It is so low that it cannot be heard more than a few yards away. It seems to be a sort of contact call, reassuring and saying, "I'm here."

But at the maras' communal dens, everybody cannot be "here." Dominance relationships are such that only one pair at a time can approach the den and call out its pups for feeding, and only one female at a time can give birth to pups there, which is always done outside the burrow. So mara den sharing has strict qualifications. The cubs retreat inside as soon as they are able after birth, and sometimes parents move their pups between dens that seem to be affiliated. Andrew concluded that increased protection for young and vigilance at the den are the best explanations for the mara's communal denning. As many as seven pairs and fourteen pups could share a den. All pups in the den are usually close in age, having been born within a week of each other, a situation reminiscent of tern and penguin crèches.

Communal denning is not uncommon in mammals, but combined with monogamy it is. The most extraordinary situation is that of the naked mole rat, which has adopted a termite-like subterranean existence with a queen, which does all the breeding, plus various workers and males. Among other species, wolves use a communal den, too, again with one female doing all the reproductive work—but in neither species does more than one female, a dominant one, have pups.

The introduced European hare, which has so complicated the predator-prey relationships on the steppes, seems a natural competitor for the mara. Might its presence account for the mara's decline in numbers? The hare also feeds on grasses and herbs, though not as selectively, and its litters are large and rapidly repeated. Gestation is only forty days, and females can conceive again immediately after giving birth. There are now so many hares that there is no doubt that they are adding to the rangeland devastation that sheep have caused. Each year, five to ten million are exported for food, mostly to Germany, without making the slightest dent in the population. However, Andrew believes that the more plausible reasons for the mara's decline are hunting and competition with sheep. Sheep seem to eat all the plants that maras seek and much else besides, and do so with exhausting thoroughness. At the same time, maras remain a target especially for sheepherders—to feed their dogs. After four of his study animals had been shot, Andrew tallied the wild game killed for dog food from August 1 to November 10, 1981, on just two of the fifty-some estancias on Peninsula Valdés: eight maras, eight guanacos, and four rheas.

Overall, the mara deserves a special place in our suite of Southern Cone animals of special concern. Like rheas, they are "canaries," very sensitive ones, even if never so widespread. The increased number of predators is probably as responsible as sheep and hunting for their obviously reduced numbers. No one has any real idea of how many are left, not even in relatively circumscribed areas such as the Peninsula Valdés. That itself is an indication of the seriousness of the situation. With diminished wildlife, the rule is, if you can't count it, you can't conserve it. Both Darwin and Musters reported maras as numerous, mentioning sightings of sixty or more animals at one time. But maras have completely vanished from areas where Bob Goelet and I found them common in the 1960s, and several Chubut estancia owners have also commented on their decrease. I fear that they are in trouble and that providing them with specially protected places may be the only responsive answer.

Maras do have a quiet constituency, one that might grow if encouraged. Many ranchers admit to liking them, if not why. Few would admit to liking the spectacular burrowing parrot, the barranquero, however, and are eager to say why.

*Our parakeets are very rapidly diminishing*
*in number; and in some districts, where twenty five years*
*ago they were plentiful, scarcely any are now to be seen . . .*
John J. Audubon

---

# 4
# *The Parrot Metropolis of Rio Negro*

Parrots are smart. Their intelligence and ability to communicate it have been the subject of so much research that they can prove it. In addition to imputed braininess, Patagonia's burrowing parrots, *barranqueros* (*Cyanoliseus patagonus*) live in spectacular colonies. The biggest of their colonies, the largest remaining parrot colony in the world, is easily visited. It is adjacent to a small resort and a large bathing beach, as well as to peregrine falcons and, a few miles away, several thousand Patagonian sea lions. As a group, parrots are the most endangered birds in the world, and 31 percent of those in Latin America are at risk of global extinction.[34] All the more reason to see Patagonia's barranqueros.

Focusing on African gray parrots, Irene Pepperberg of the University of Arizona has found astonishing abilities in communication and perception — in the birds' ability to speak their minds. Concluding her 1999 book, *The Alex Studies*, Pepperberg summarized some of the reasons she has struggled to "speak" with parrots:

> If the data help us respect the processing abilities of brains that are structured differently from those of humans or are used to better the life of even a single captive parrot, prevent habitat destruction and capture of birds in the wild . . . my work, . . . and the work of my parrots, will not have been in vain. . . .[35]

On the jacket of Pepperberg's book, Harvard scientist Donald Griffin succinctly summarized her research:

> Thorough and critical investigations have led to the truly revolutionary discovery that the parrot can literally mean what it says . . . [and] demonstrate that we can gather significant evidence about nonhuman thoughts and feelings by adapting the same basic approach we use with our human companions: namely, to let them tell us what they think and feel.

Patagonia's barranquero deserves to be taken seriously, and not only for its probable intelligence. Its spectacularly coordinated flights and cliff-face cities are what won my admiration. The first wild barranqueros I ever saw were near Carmen de Patagones in the province of Rio Negro, where Musters completed his extraordinary adventure and which, in Darwin's day, was "the most southern position (Lat 41°) on the eastern coast of America, inhabited by civilized man."[36] In 1964, at the time of my first visit, Bob Goelet, Bill Drury, Kix, and I were putting together our Argentine wildlife movie, and Carmen, the gateway to Patagonia, was a fitting locale to film.

In the town of Carmen de Patagones, buildings and walls were decorated with modern portraits and graffiti lauding Che Guevara, which appeared behind antiquated oxcarts trundling through the somnolent streets and past an imposing old church. The latter's kindly caretaker showed us about and took us to his quarters to see his pet barranquero —with which he exchanged gentle chuckles and rubbed noses. We purchased a few supplies, had our picture taken tourist fashion in the town square, and then visited a former New Yorker named Trousdell who was ranching on the south shore of the Rio Negro just east of Viedma. Mr. Trousdell, a big, powerful man who wore the practical local garb of *boina* (beret), *bombachas* (baggy trousers), and rope-soled *alpargatas* (shoes), confirmed our choice of hotel and indicated a safe local restaurant. He then took us into his big backyard to watch barranqueros by the hundreds fly in to roost for the night.

As a dust-reddened sun gradually sank out of sight, the parrots gathered—gaudy, screaming flocks assembling from northern Patagonia's measureless stretches of dry scrubland. From the accounts of friends in Chile, we had expected the bird to be rare. The slightly different Chilean race had been persecuted almost out of existence (only thirty-three hundred were left in 1985),[37] so we marveled at Mr. Trousdell's flocks. About eighteen inches long, the barranquero is mostly olive, with shimmering blue wings, a yellow rump, and a breast banded with scarlet. The birds clustered in the tops of a long row of Lombardy poplars and squabbled over favored perches. Many roost in the poplars all night, but Mr. Trousdell remarked that others fly across the river to the high sandstone cliffs on the north bank, where they nest. As it grew dark, the barranqueros became quiet silhouettes, and we joined Mr. and Mrs. Trousdell and their son for a glass of strong red wine and a peck of conversation.

Parrot calls were clearly audible as the birds flew over our hotel at 3:00 a.m. We loaded the station wagon and got under way while the town was still asleep, taking a road behind the cliffs we had spotted from the Trousdells'. Martinetas dashed across our path, while dusky Patagonian mockingbirds sang from fence posts. A mara bounced along beside us with a fox in half-hearted pursuit. It was a crisp, cool morning, and we found a track leading down toward the Rio Negro. The owners of a riverside farm gave us permission to park and carry our camera gear through their vegetable gardens to the high barrancas where we hoped to find nests.

Were there barranqueros? "Si, hay muchos!" said a farmhand. The low cliff face was perforated with holes about four inches in diameter, and a startled flock of two dozen birds flew up upon our arrival. The burrows were said to be three to five feet deep, and some were so low that I could just reach them. The sandstone in which the parrots burrow is soft as rocks go but hard enough that the burrows represent prolonged mining. I saw a bird digging, and it seemed to be using both feet and beak. Bob and Bill traipsed farther along the riverbank toward the tall hills nearer the Rio, while Kix and I arranged our cameras and tripods under a small willow within a few feet of the first cliff and sat down to await the parrots' return.

Years ago, barranqueros wintered in immense flocks as far north as Buenos Aires. Today they are scarce there, although still common in a few places. Almost everywhere, however, they are shot as crop pests—easily. They show great distress when a flock member is killed and circle over their fallen companion, thus offering the shooter repeated opportunities, a behavior identical to that of the United States' now extinct Carolina parakeet—reminding us of the roots of altruism. They were cautious when approaching the nest burrows in our presence, though. First, they pitched into a tree nearby and watched us. Then the flock rose and flew screaming past the face of the cliff and out of sight, suddenly reappearing fifty feet over the edge of the barranca, soaring for twenty, thirty, forty seconds at time in a compact formation on the updrafts at the edge of the cliff. It was a lovely sight and a surprising one. We didn't expect parrots to soar.

Eventually, two suddenly flew off, to reappear in a rush at the cliff face and vanish into their nesting burrow before I could begin to locate them in the camera viewer. Such speed and precision flying frustrated photography, and I spent several hours attempting to obtain portraits of the birds at a nest entrance. By the time we were ready to leave, however, they had become so used to us that they merely chattered as we tried to walk inconspicuously away along the foot of the cliffs. We didn't realize it at the time, and none of the townspeople we spoke with thought it of note, but just a few miles farther on was the most spectacular barranquero colony in the world. It would be years before I returned to explore further.

At the northeast corner of Patagonia, near where the Rio Negro flows into the Atlantic, the sandstone cliffs are more than one hundred feet high. There, immediately beside the resort village of Balineario El Cóndor, is a magnificent parrot city that borders the beach and the sea for 7.5 kilometers (4.7 miles). The nests honeycomb the tall cliffs. In the first kilometer, there are nearly seven thousand. It is dazzling.

Wheeling, diving, soaring, gliding, the olive parrots flash by the cliff in flocks of hundreds, like fighter planes painted with squadron symbols of yellow, red, and blue. Pairs suddenly peel off or approach the vast cliff city on their own and shoot into their burrows like bullets. They also loiter on ledges near the burrows, preening, chattering, and feeding their mates—a restless psittacine embroidery on a city of stone. Southern martins (*Progne elegans*), midnight black with purple gleams, also nest on the cliffs, using old parrot burrows, and we watched a chimango hawk (*Milvago chimango*) discreetly investigating one burrow after another, seeking an opportunity to catch a chick near the entrance—all the while being badgered by martins and vigilant parrots. Peregrine falcons also nest on the cliffs, but they seemed to be ignored by the parrots.

Farther to the south, as I watched barranqueros soaring over the four thousand sea lions congregated at Punta Bermeja on the rocks below the cliffs, it occurred to me that, like so many other colonial creatures in this curious land, these birds are also island nesters. Instead of breeding on water-surrounded islands, as do most seabirds, or on the make-do islands of isolated puntas or the flamingos' muddy islands in isolated salt lakes on the steppe or altiplano, the barranqueros breed on cliff islands, whose inaccessibility provides the same protection from predators that the other colonialists find on conventional islands. To preserve any of these creatures, we must save their islands.

Along the road to El Cóndor, small flocks of barranqueros sipped fresh water from roadside pools and nibbled white, salty mud. Others foraged at the feet of big, stolid Herefords and perched in conversational rows on telephone lines. Their interchanges were boisterous and seemed full of opinion. Recently, some sophisticated science conducted by Juan Masello and his colleagues from the Institut für Ökologie in Jena has given us additional insights.[38] Among other things, it addressed controversial family relationships.

For more than a century, ornithologists have claimed that burrowing parrots "mate for life."[39] This is a bold assertion. As maras remind us, strict monogamy in wild animals is rare. Although pairing for a breeding season is usual among birds, most pairs are not as faithful as they might be. In some species, more than half the chicks are of extra-pair origin, although a high level of faithfulness seems to be the rule among albatrosses, penguins, owls, and birds of prey.

In early December 2003, Juan Masello was rappelling down the gray El Cóndor parrot cliffs in an uncomfortable-looking harness, dangling precariously in the cold wind from a hundred-foot rope. Below, his scientist wife, Petra Quillfeldt, waited with two colleagues, boxes of chemical reagents, scales, calipers, and dozens of tiny vials. Jamie Gilardi from California, executive director of the World Parrot Trust, was there, too, with Graham, Kix, and me. We had gathered to talk about saving the parrot colony. Juan's work had called attention to the uncertain future of the birds when he decided to address the barranquero fidelity question (among others) with a DNA fingerprinting study.[40] Taking tiny blood samples from the chicks, which Juan gently pulled from their nests and quickly replaced, Juan and Petra tested forty-nine barranquero families, eleven of them for two years. Their findings confirmed the old claim of barranquero faithfulness. Despite the closeness of living conditions, some nest burrows actually interconnecting, not a single "extra-pair paternity" was found. Masello makes the point that such strict monogamy occurs primarily in birds such as long-lived parrots with long reproductive life spans and where paternal care is essential to successful chick rearing.

Feeding, nesting, roosting, traveling together, the barranqueros' close-knit families and societies are almost an exaggeration of the lives of many other social creatures. For such small creatures, they also live a long time, far longer than a dog or cat; twenty years and more is not unusual for parrots of this size. Even after their chicks fledge, they care for them outside the nest for an additional four months. From what we know about parrots in general, and now about barranqueros in particular, it seems reasonable to conclude that they are exceptionally responsive, have good memories for individuals and places, and mate faithfully, perhaps even for life. Juan says that breeding barranqueros stay in their nests at night with their chicks, that parents feed chicks four to five times a day, and that both parents usually arrive at the nest to feed their chicks at once, so pairs probably forage together. These are very affectionate birds.

In 2003, Juan and his colleagues counted thirty-five thousand breeding barranquero pairs in the long El Cóndor nesting cliffs. But they also compared the number of active nest sites with that of previous years and discoverd a dramatic decline. North America's Carolina parakeet (*Conuropsis carolinensis*) offers a depressing precedent. Although it was at least as abundant as the burrowing parrot, quite beautiful, and ranged from New York to Florida to Colorado, it is now extinct. The last wild birds disappeared about 1904. The barranquero, half again as long as the Carolina parakeet and much more spectacular in its gatherings, is beginning to be threatened at its most major nesting area, and few seem to care. At El Cóndor, there is not a single sign boasting of their presence or the unique nature of this, the world's largest remaining parrot colony. Not one ranger protects them—even though protection could hardly be simpler, and even though announcements of their beauty and their presence from September through January would bring tourists in droves. Instead, they are being deliberately killed and collected for sale, and the process is being accelerated by ignorance and cruelty.

Rio Negro colleagues told us that besides being captured for the pet trade, the birds are being poisoned as pests. In 1984, the barranquero was officially declared a crop pest under Argentine law. Poisoning, already widespread, accelerated. In 1995, such poisoning was stopped officially but not actually. The reprieve took place not because of concern about the birds but because poisoning was perceived as dangerous to humans and livestock. Meanwhile, it goes on, and capture for the pet trade continues. In 1992–1993, the export quota was 16,200. When we visited in 2003, a Punta Bermeja ranger told us that an organized group that arrived with trucks had taken some 4,000 birds from the cliff in the previous few months. Recently, the beach below the El Cóndor nesting cliffs was opened to ORVs and 4x4s—unnecessary to access the convenient beach. We found it crowded with vehicles, and Masello's team reports that the trucks on the beach so frighten the parrots that they are often unable to feed their chicks.

During another visit, we happened upon a pitiable little team of parrot collectors—not all are organized commercial animal traders. It consisted of an obviously poor, middle-aged woman with three little girls and a small dog. They were equipped with hooked poles and a basket of sacks. Their collecting strategy consisted of sending the most agile of the children up the cliff to a reachable nest to pull out chicks. They had five. One already had been injured. The chicks were frightened, blinking in the light, still flightless, and trying to hide. Their plight and what I judge was that of the collectors made me both angry and sad. As a biologist who has bred parrots, I asked the woman what she would feed the birds. She explained that she mixes bread and water and gives them some sunflower seeds—a prescription for malnutrition. She had been collecting the birds for some years, admitted to losing "some," and added that she was not the only collector and that some people were shooting them.

How much damage do the barranqueros of El Cóndor do to nearby crops? How are export quotas set? Who obtains permits and with what qualifications? For the most part, these are unknowns. But the perception of the barranquero as a crop pest and the general lack of interest in wildlife are central to its straits.

In 1997, while we were again watching barranqueros at El Cóndor, Graham, Kix, and I met with Cecilia Vinci, who had worked with us on a coastal management plan. She listened to our concerns about the parrots and arranged for us to meet with Rio Negro governor Raul Verani to talk about this and other issues.

Sr. Verani was not, it turned out, a happy man the day we saw him. Red-eyed (he had been up all night at a political meeting), sick, and distracted, the *gobernador* nevertheless summoned up that inherent resource of courtesy that makes meetings with most Argentine high officials pleasant, whatever the situation. To this, he added coffee and as attentive an ear as his condition permitted. When it came to barranqueros, I provided some on-the-spot (and rather imaginative) parrot math to show that barranquero damage to agriculture is exaggerated —which surprised the governor and his staff but has not had much effect.

For example, burrowing parrots weigh about 9.5 ounces each. Let's assume that there are seventy thousand at the El Cóndor barrancas and that they eat 5 percent of their body weight each day (it is probably between 3 percent and 4 percent). So, El Cóndor's parrots could weigh about 41,563 pounds altogether and eat 2,078 pounds of food a day when they are in residence. I think it can be shown that the vast majority of their food comes from wild bushes and shrubs, but, for this calculation, let's say that 30 percent of it comes from farmers' crops and, also, that the birds destroy as much as they eat (60% × 2,078 = 1,247 lbs. a day). However, most of the birds don't feed in this area for much of the year, and the crops don't grow year-round and aren't vulnerable all the time. Let's charge the parrots for five months of food and destruction, which means that we should subtract seven-twelfths of 1,247 pounds for the time they are not here. That still leaves them with an annualized daily liability of 727 pounds, which adds up to about 133 tons of crops a year. While I have deliberately overstated what I think the birds' consumption is, 133 tons sounds like a lot. Are they worth it? How can we provide perspective to the gobernador and the farmers? How, for example, does this compare to the food eaten by the Hereford cows we saw?

Those big cows weigh about 900 pounds each and eat at least 3 percent of their body weight each day, 27 pounds of food. Thus, seventy thousand parrots at El Cóndor might be eating and destroying the equivalent in crops, in this region, of what it takes to annually feed twenty-seven cows, about 133 tons. This is not science, of course; it is speculation. A good scientific study would correct these figures and accurately address the sources of food that parrots and cows are eating, as well as pasturage, veterinary care, shipping and "finishing" for cows, and many other elements that I have overlooked, not to mention the farmer's income. He gets none from the parrots. Nevertheless, this speculation provides some basis for asking whether a natural treasure is worth as much as twenty-seven cows. Even a little advertising and promotion of the parrots and viewing opportunities could produce far more income than the cows.

In 2003, Juan, Graham, Jamie, and I wrestled with parrot preservation ideas. A local group was working on signs, Juan said, and Parques Nacionales had been brought to accept the idea of a national park—but had done nothing. Worse, one so-called conservation group was pushing regular "harvesting" of the birds for trade. Such an idea can be advanced only by people who are unfamiliar with the miserable conditions and corruption inherent in so much of the wild parrot trade. (If people want barranqueros for pets, let them breed them in captivity. Fewer will die in misery; the birds will be happier, better pets; and the great parrot city will still be there for everyone to enjoy in untrammeled magnificence.) We came up with the outline of a plan, and, fortuitously, Graham and I were able to speak to Dr. Miguel Saiz, the new governor of Rio Negro, a few days later. He seemed sincerely interested, and an international conference on the situation is now planned for 2005.

Things can be done: Promote the great El Cóndor barranquero metropolis along with Rio Negro's other wonders. Put a warden on the site. Add interpretive signs. If necessary, pay a fair compensation to those farmers who can prove serious parrot depredation. And don't worry too much about my admittedly desperate shirt-cuff math. In the long run, barranqueros will be conserved only if people value the birds not just for pesos, but for themselves, as part of the region's unique environment, one of the greatest wildlife sights anywhere, a part of the living lineage of Patagonia.

*Grown-ups never understand anything for themselves,*
*and it is tiresome for children to be always and*
*forever explaining things to them.*
Antoine de Saint-Exupéry

---

# 5
# *The Tortoise and the Bus Driver*

Few living lineages compare with that of a tortoise, but a tortoise in the Patagonian desert, plodding along under dashing flocks of barranqueros, ignored by haughty guanacos, and, if very small, eaten by choiques, is hardly among Patagonia's greatest wildlife sights. Few people ever see one, but it is the Southern Cone's biggest terrestrial reptile (which isn't saying much), fascinatingly out of place (no other tortoises are so far south). It is an intrinsically interesting part of the region's diversity, and its future is increasingly uncertain.

Tortoises have a special cachet. There's the one that raced the hare and the one said to have killed the ancient Greek historian Herodotus when dropped on his head by an eagle, and certainly the giant tortoises that so astonished Darwin and gave their name to the Galapagos Islands. But the tortoise I remember best was a Patagonian in the possession of a bus driver in Puerto Madryn. It had attracted a small crowd.

A dusty little vortex of people was circling in the wind and sand off the walk across from our hotel. Its center was a dark, wiry fellow in the unmistakable blue shirt and shiny serge pants of a bus driver, and its shifting windrows were mostly small boys. The eye of the storm, somewhere about their feet, was an almost washtub-sized tortoise. It looked anxious, with as much as possible of its considerable body and head withdrawn under its capacious carapace and behind its armored forelegs. Those thick, strong legs were covered with close-set spiny scales like cut-off porcupine quills and must have discouraged many a fox and pampas cat. But now it had reason to worry more, for the driver and the boys were exerting themselves in the way humans often do when confronted by a tortoise: trying to get him to move.

From a tortoise's prolonged point of view, from a family tree that dates to the Triassic, pokes and proddings from creatures whose recognizability scarcely extends 200,000 years could hardly be anything new. Its kind had stoically outwaited similar antics from a panoply of creatures back to the days of the dinosaurs—and most of the pokers and prodders are gone.

Having wormed my way into the midst of the onlookers, I gingerly assumed the vulnerable position necessary to examine the posterior of a tortoise and, levering it up a bit, saw a flat, not concave plastron and short puny tail. Graham duly translated my diagnosis that this was a lady tortoise to the onlookers, and the chattering turned from jokes to natural history—and an interrogation of the bus driver began.

This tortoise was a big, old animal, dignified in the way old tortoises can be—their wrinkles and confidence reminding you that, with any luck, they will be here long after you are gone. I had visions of the famous tortoise that dwelt on St. Helena with Napoleon but made no trouble and lived 177 years, and of Lonesome George, the solitary remaining Pinta Island Galapagos tortoise living at the Charles Darwin Research Station on Santa Cruz Island, all its mates killed by sailors, its food and young done in by goats, pigs, and rats introduced to the island. This Patagonian tortoise's luck had also run out. With the driver's permission, I measured it—carapace 15.3 inches long and 9.5 inches wide, straight, not on the curve, much bigger than others of its kind I had seen in the Buenos Aires pet shops and zoo. It was certainly aged but also healthy enough to regard its ring of admirers with apprehension, especially when they diverted its persistent attempts to depart. It was a Chaco tortoise, scientifically named *Geochelone chilensis donosobarrosi*—presumably against its will and not even in agreement by all the people who name tortoises. Some think it is not different enough from the rest of its species to merit a subspecific name. The species is found all the way north to Paraguay and Bolivia and is thought to be the closest living relative of the giant Galapagos tortoise. That indicates a species split that probably occurred 6 to 12 million years ago—about the same time our ancestors separated from those who became chimpanzees, just yesterday for a tortoise but a long time before buses. The driver had obtained the tortoise along the road on his way to Puerto Madryn and was bringing it home as a pet for his children.

Five years before, Graham, Kix, and I had found a much smaller Chaco tortoise on the road to Punta Ninfas, just south of Puerto Madryn but far south of the species' known range. I remembered mention of one even farther south, in one of Gerald Durrell's marvelous books. Could it be that this was a cold weather tortoise? I later spoke to Lee Durrell, Gerry's wife, and received a quick response from Gerry:

> Lee tells me that you have been reading bad literature, i.e., one of my books, and have come across a reference to a tortoise. I can't be terribly specific as to the exact spot where I saw this tortoise but it was well below Puerto Deseado, about half a day's drive down the coast towards the big penguin colony there.

Gerry, then, had found his tortoise, certainly another Chaco, roughly 500 miles south of its southernmost range and 350 miles south of Puerto Madryn. We soon learned that neither the area's biologists nor its farmers had ever seen or heard of a local tortoise. So Gerry's find was the victim of a tortoise pet traffic, like that belonging to the bus driver. The Chaco tortoise does not go anywhere near Puerto Deseado or even Puerto Madryn. As we talked with the young bus driver, a curious story emerged.

The driver's route took him through Rio Negro province, immediately north of Chubut and Puerto Madryn. He found he could buy tortoises at a canal keeper's station and sell them later as pets. This tortoise, though, was so big that the driver considered it special and was bringing it home.

"Millicent," as I privately thought of her, had the rear edges of her carapace handsomely flared, and the broad scutes on her back each had a dark center. It was her imperious face, however, that won her my maternal grandmother's middle name. Higher and narrower than the faces of most tortoises, Millicent's held dark, clear eyes and an overall expression of experienced reserve. She was an impressive creature, but her future, like that of most wild pets, was bleak.

Studies published in 1987 estimated that seventy-five thousand Chaco tortoises were sold each year as pets in Argentina, and that 32 percent died within one year. About three thousand more are exported to Chile, Denmark, Germany, Japan, the Netherlands, Uruguay, and the United States.[41] John Behler, the WCS curator of herpetology, tells me that the species became relatively common on reptile dealer price lists some time ago and continues to be offered from time to time. About one-third of the world's three hundred species of turtles and tortoises are now endangered, he adds, especially in Asia, where they are eaten in vast quantities. At least 2.6 million are sold each year in China for food, and they have become extinct over large areas.[42] Americans are also guilty in this regrettable trade. Between 1989 and 1997, 5,709,972 turtles and tortoises were imported, of which 729 were Chaco tortoises. The total number of exports in the same period was 53,751,521, mostly U.S. red-eared turtles (*Trachemys scripta elegans*) but also 79,122 slow-breeding box turtles (*Terrapene carolina*).[43]

Later, I tried to buy Millicent from the driver—unsuccessfully at first—so that she could be released. In the meantime, Graham questioned other drivers, and it became clear that a number of bus and truck drivers had developed an irregular (and illegal) business in "pet" tortoises. This is not a common animal, and it was already listed as threatened on international conservation lists. Where, then, was the canal keeper getting his supply? We were directed toward Rio Negro province and the seventy-seven-mile Canal Pomona–San Antonio Oeste, and we decided to investigate.

The canal was not hard to find. It was twenty feet or more wide at the top, with sharply sloping concrete walls, and it ran through largely uninhabited shrub desert. We startled a group of three martinetas and a small flock of chingolos (the pretty little native sparrows) that were foraging along one bank of the first section we saw, and a chimango hawk followed us as we jumped the sheep fence that guarded access to the canal. Water depth varied from less than a yard to a good deal more, but the flow was slow. A short walk turned up three tortoises struggling despondently in the water and trying to climb the steep concrete wall. We found one dead and one lying exhausted at a weir meant to catch debris. The right foreleg of the exhausted tortoise, a male, was severely burned—fires set in the shrubland to make more suitable pasture for sheep or cattle are another human-made hazard for this beleaguered creature. Soon we found more carcasses near the canal's weirs, where the keeper had pulled out dead ones. None approached Millicent in size.

Nothing in the Chaco tortoise's long experience of the world, an experience that has included mylodons, megatheriums, glyptodonts, sabre-tooth smilodons, and pre-guanaco hemiauchenias, as well as sheep and fire, had prepared it for an artificial canal across the middle of its ancestral home. Formed of smooth concrete with sloping sides that drop eight to ten feet into the water, the canal's design constitutes a death trap for tortoises, which can rarely climb out. Although access to the canal was guarded on each side by a sheep fence, even the largest tortoises could easily walk under its lowest strand.

A barrier far lower than the canal sides, the concrete curbs along the streets of suburban Westchester County in New York, and in much of the rest of the United States, creates just as formidable an obstacle to the free passage of North American box turtles and salamanders and leads to the deaths of thousands of turtles and amphibians in the country each year.

Marcelo Guerra, a canal attendant, told us that he removes twenty to thirty tortoises each month from the dams and weirs in the section for which he is responsible. The weirs are set at intervals along the canal to control water levels and trash. Tortoises are most abundant in the canal in December, he said, which probably coincides with their breeding season. He sells live ones to passing drivers. Dead tortoises are burned on the shore with canal debris raked from the weirs. The attendant at another canal station, Sr. Ziede, added that guanacos, Darwin's rheas, pumas, and wild boar also drown in the canal and he must get their bodies out. With the help of Sr. Ziede's three children, we rescued two more young tortoises, and the children walked with us into the nearby desert to release them.

In San Antonio Oeste, supplied by the canal water, we found that the plight of the tortoises was well known and that a sympathetic campaign had been started by schoolchildren. Saint-Exupéry would not have been surprised. Eighteen inches of wire mesh (turned out two inches at bottom) placed at the soil line along the lowest part of the canal's livestock fences would correct the problem, but whether anything will be done along those lines remains to be seen.

It would be hard to make a case for the tortoise as a landscape or umbrella species, a creature whose conservation would be likely to provide for the well-being of many other species. Compared with the guanaco or Darwin's rhea, its range is circumscribed. You must get down on your hands and knees to consider it charismatic in the usual ways, but if schoolchildren have already identified with its plight, its charisma may be underrated. As a key species, it deserves another look. Perhaps for its specialness as the biggest reptile in Patagonia and the venerability of its history, it deserves a mini- "Triassic Park" of its own.

Chelonians—turtles, terrapins, and tortoises—are conservative. Most cannot be mistaken for anything else, and tortoises are the most conservative of the lot. Three hundred million years of experience fairly shines out of their simple forbearing ways. I had the good fortune to grow up in the outskirts of St. Louis, Missouri, where the attractive three-toed box turtles were then common. I fell in love with them as only a child discovering such a wonderfully colorful creature in his own backyard could. So I studied them, photographed them, bred them, and wrote my first zoology paper about them. They stimulated my first serious thinking about nature, and, to this day, I particularly hate to see a turtle killed, whether on purpose or through ignorance. So what I say about tortoises may not be entirely objective.

In Puerto Madryn, I put up $20, and Graham negotiated. After a few days, a note in an unfamiliar hand appeared in my mailbox at the hotel. Swayed by the eloquence and principle of Graham's arguments, the bus driver's wife wrote, they had decided the tortoise was to be ours. She was happy we had called the tortoise's importance to her husband's attention. And we could have the animal quite simply for the welfare of that tortoise and its kind and with her good wishes—for $25. Graham and his children, Edward and Sabrina, returned Millicent to the home site the driver described, and he and I began puzzling how to add tortoises to more advanced conservation plans for coastal wildlife such as seals and penguins. I fear that only a reserve and canal fencing will save them.

Meanwhile, just southwest of the Chaco tortoise's largest concentration, a dramatic experiment in wildlife restoration is taking place.

*An election is coming. Universal peace is declared,*
*and the foxes have a sincere interest in*
*prolonging the lives of the poultry.*
George Eliot

---

# 6
# *The Condor, the Puma, the Fox, and the Aliens*

The hot sun burned down upon Patagonia's parched, sparsely populated Somuncurá plateau in south central Rio Negro province, as usual, but this morning it warmed a straggle of diverse and companionable marchers. It was December 22, 2003, and more than 450 men, women, and children were traipsing on a faint stony path toward Las Sierras de Paileman, a forbidding line of black basalt massifs whose height dominates the desert hereabouts. The marchers were festive and carried big beautiful blue and white Argentine flags. They had come to witness an attempt to bring back to its old home on the steppes the great Andean condor (*Vultur gryphus*), a creature admired everywhere as a symbol of unsurpassed flying skills, one whose name is borrowed for high cliffs, athletic teams, and military squadrons. It is among the largest and most legendary of all flying birds.

We mostly see condors in the wild as big silhouettes circling in a distant sky, their improbably broad wings widely spread and unmoving as they swoop effortlessly over high mountain ridges. Their nearly flapless progress fascinated Darwin:

> When the condors are wheeling in a flock round and round any spot, their flight is beautiful. Except when rising from the ground, I do not recollect ever having seen one of these birds flap its wings. If the bird wished to descend, the wings were for a moment collapsed; and when again expanded with an altered inclination, the momentum gained by the rapid descent seemed to urge the bird upward with the even and steady movement of a paper kite. However this may be, it is truly wonderful and beautiful to see so great a bird, hour after hour, without any apparent exertion, wheeling and gliding over mountain and river.[44]

But condors feed on death. A circling group usually means that they have found it, and the approach of such huge birds can be a bit unsettling. High in condor country, on Neuquén's Auca Mahuida, "wild mountain" in Tehuelche, Graham stretched out on a large, flat rock and did his best decomposing-carcass imitation—no task is too risky for the dedicated. Within minutes, seven condors, which had been circling something tasty a few kilometers away, were eyeing him, and one landed nearby. The only birds that came close were immatures, however. The adults, easily distinguished from brownish youngsters by white neck ruffs and black feathers, were not fooled.

Only the long and narrow pinions of the wandering albatross (eleven to twelve feet) exceed the spread of the condor's huge wings, which can reach ten feet, but condor wings are far wider. Yet, as weight increases, the requirements of flight outpace the capacities of muscle power. Thus, larger birds, whose lifestyle requires them to spend much time aloft, must make greater use of soaring and gliding than smaller ones, and their maximum size is limited. Until thirteen thousand years ago, about the time the first humans reached Patagonia, condors shared the sky with such huge, condorlike avians as *Teratornis incredibilis*, which had a wingspread of nearly fifteen feet, and *Argentavis magnificens*, which weighed up to 80 kilograms (176 pounds) and had a wingspan of thirty-six feet, the size of a small Cessna.[45] Given what we now know about the relation of muscle power to wingspread, it is hard to see how they flew at all. Male Andean condors and wandering albatrosses weigh about the same, 25 to 28 pounds, and females of both species are noticeably smaller. However, softly feathered albatrosses and fierce-featured condors look nothing alike. The huge, bald, red-skinned head of an adult male condor is so wrinkled, wattled, and ugly that it ends up looking dignified—and these birds, again like albatrosses, can become very old. Some have lived more than sixty-five years in zoos.

Although "tame condor" seems an oxymoron, one reared at the Frankfurt Zoo had the run of the city. I was shocked to see her swoop down, land on the sidewalk, and make greeting gestures to a pair of apprehensive strollers. For some months, she was a city institution—until she discovered what fun it was to tramp around on the hoods of shiny Mercedes and pull off rubber windshield wipers with her powerful beak. And what a beak—as a big male at the St. Louis Zoo taught me one day by reaching over a pan of food I was offering and grasping my left thumb. I retained that cherished digit, scarred though it is, after an unseemly struggle before a delighted group of zoo goers. He reminded me of the condor's ecological role in opening tough-skinned carcasses that smaller carrion eaters, such as foxes and caracaras, couldn't easily tear. We had long been friends, and, with regard to my thumb, I'd like to think that he made an honest mistake.

Substantial numbers of condors still breed in the Argentine and Chilean Andes, but the big bird became extinct in Venezuela in 1965, its populations in Peru and Bolivia are greatly reduced, and less than one hundred birds survive in Colombia and Ecuador. All of the populations that once ranged out to the Atlantic coast over the Patagonian steppes are gone. To bring them back is the long-time dream of the imaginative Luis Jácome, general curator of the Buenos Aires Zoo and an old friend. At last, in the vast Somuncurá in Rio Negro province, Luis had found a promising place to try a restoration on the steppes. The Somuncurá is a scantily populated tableland of 15,000 square kilometers (3.7 million acres), 1,000 to 1,500 meters (4,900 feet) above the surrounding steppe, with peaks up to 2,000 meters (6,500 feet) and spectacular rock formations. It was declared a natural reserve in 1986.[46]

Realizing that condor reintroduction could be successful only if the people who live in the region supported it, Luis spent months getting to know Somuncurá residents, most of whom graze sheep and cattle in the rocky brushland and are of Mapuche Indian descent. Eventually, Luis won their interest and support and arranged to release the birds on Las Sierras de Paileman. One of the oldest residents remembered his grandfather speaking of condors there, and he described them recognizably. The spot was close to where Darwin had seen condors 170 years earlier, Luis pointed out. By coincidence, Darwin had seen his birds at El Cóndor, the same towering parrot-city cliffs where we had met with Juan Masello and just east of where Millicent and her Chaco tortoise kin are living.

The morning condor march included virtually every local person: the provincial governor, the news services, wildlife officials, Somuncurá families, and guests like Graham, Kix, and me. We assembled at the simple adobe ranch house of our handsome Mapuche host, Sr. Botana, where the best parts of a young bull, which had been slaughtered for the event, hung from racks under thick-trunked willows. A group of elderly local matrons, using the occasion to exchange news, formed a colorful welcome, sitting on benches and boxes in front of the house. Across from them, parked helter-skelter in the bush, was a diverse assemblage of ranch vehicles, some of which looked as though they might also remember condors. On word that Luis and his family were arriving, flags were unfurled, and we began the three-kilometer hike to the looming massif.

Five young condors, which Luis had raised for this experiment, had been kept atop the Paileman cliffs for two months of acclimatization in a mesh aviary and wooden shelter designed to minimize contact with caretakers. A cluster of tents in the brushland below housed an enthusiastic group of volunteers from the Buenos Aires Zoo and the Bio-Andina Foundation. When Graham, Kix, and I arrived, the volunteers had been camping there for weeks to care for the condors and determine how to follow them once released. Below the cliff, where we stood, the volunteers had built a long shelter for onlookers, so that the marchers could watch the release without disturbing the birds on the big day.

The release began with a short song of hope in Mapuche by the community's elderly representative, Lonco Manuel Cayú, from high atop the condor cliff, inviting the condor spirit to return to this place. Then, the five sons of the family Botana, owners of the property, emptied a basket of twenty-inch (fifty centimeter) molted condor feathers that, caught by the desert winds, swirled and floated slowly down toward the watchers far below. Finally, the door to the shelter was opened, and the five young condors (each about two years old) stalked calmly, quizzically, out of their enclosure (nearby newspeople and keepers were hidden in clifftop blinds). They looked about for a few minutes, carefully, it seemed to me, and then three of them suddenly leaped into the wind and were soon soaring high above the cliff and our heads as everyone, children too, held their breath—a moving and remarkable occasion. After all, these birds had never soared or really flown before; they were much too big to do that in even the biggest zoo aviaries. They circled for a while in the Patagonian gusts, marvelously and obviously born to live in them, and then returned. All had radios and were to be tracked and brought back to the release site if they got lost. Food would be provided there, so they had to learn where emergency rations could be found. NASA and ARGOS satellites were helping to make the condor tracking possible.

With this sort of public involvement, the likelihood that the birds might immediately be shot is reduced. Luis has observed, "It takes three years to release a condor, but only seconds to kill it." But there are other worries. Although the Somuncurá is an area largely uninhabited by people, it is not clear whether there is enough food to support the condor population. Still, guanacos, dead sheep, cows, and horses could provide food, as they now do in the Andes.

Reintroductions almost always suffer high mortality and require a great many animals until viable populations are established. More than two thousand peregrines were lost (many to great horned owls) in their eventually successful reintroduction in the United States following the DDT catastrophe, which had destroyed most of their populations. However, the condor is a different kind of creature—a markedly thoughtful one. Mike Wallace, a condor expert at the San Diego Zoo who is trying to reintroduce California condors to Baja California, thinks they have a chance. At the least, the increased interest in wildlife restoration that this effort creates is useful, and Luis has demonstrated the importance of community involvement and public relations for such a complex reintroduction wherever it occurs. Successful or not, the implications of the experiment are fascinating. They encourage our thinking about the future of the steppe's wildlife and about the wonderful wild pageant that existed only a little while ago in the days of the Tehuelche. But for now, who or what will set the condor's table?

Condors are not killers but scavengers. Although they may kill an occasional newborn chulengo, sheep, or calf if it is weak, sick, or poorly protected, stories that they fly off with sizable lambs, calves, and even children are just that: stories. Could a condor carry prey in its beak to its chick? Yes, theoretically, and there are so many reports of condors making off with very small newborn lambs (more likely a piece thereof) that I cannot entirely discount them.[47] But a condor's feet are more like those of a turkey than an eagle. I have not been able to find a single creditable account of a condor carrying anything in its feet—despite one unforgettable cartoon of an auto parked on a mountain road, a condor carrying off the driver in his claws, and the driver's wife screaming up, "Drop the keys!"

Generally, condors eat big dead animals. They tear off and swallow pieces and regurgitate the mess for their chicks, like wolves, Cape hunting dogs, hyenas, and vultures (although condors are thought to be more closely related to storks than to African vultures). In contrast, eagles kill their own food and bring it to their nestlings whole or in large pieces in the talons of their powerful killing feet. Until humans came along, condors were dependent on some sizable creature's bad luck, which, since the demise of the ancient Patagonian horses, megatheriums, and such, would have to have been guanacos, vicuñas, rheas, maras, probably flamingos, swans, and geese on the steppes and altiplano, and, on the Atlantic coast, a variety of marine birds and mammals. Once the Pleistocene predators were gone, the table setters on the steppe and altiplano, besides human hunters, can only have been pumas and the big foxes called culpeos for those meals that did not die "naturally."[48]

On the Pacific coast, condors still feed upon dead seabirds and marine mammals that have washed up on the shore, and also in seabird colonies. They must have done the same thing on the Atlantic shore, and the high cliffs along this shoreline might have supported nesting in the days of the Tehuelches. With millions of guanacos and rheas and thousands of pumas and culpeos setting the table with them in presettlement times, there should have been plenty of condor food and hundreds of condors on the steppes, as well as the altiplano.

Today, the condors are gone except in the mountains, leaving carcass scavenging to modest turkey vultures, hawk-like caracaras, foxes, and armadillos. Whether the condor can live again on the steppes may depend on the availability of carcasses and the restraint of wanton shooting. Its world has been dramatically altered by alien species, and its reintroduction highlights the fact that present reserves are badly managed and inadequate for steppeland animals.

For present-day predators, such as those pumas and culpeos, the abundant introduced sheep, European hare, rabbits, and red deer are much more important prey than the diminished populations of native animals. The hare has become an almost unfailing source of food for them and for smaller gray foxes and birds of prey. Over 58 percent of the food of Patagonia's big, black-chested eagle buzzards (*Geranoaetus melanoleucus*), whose population density has become the highest known for large eagles, now consists of European hares.[49] In the province of Neuquén, 94 percent of available prey biomass is now domestic or exotic, mostly sheep, hares, and rabbits. But rather than diverting predation from the now scarce and beleaguered native animals, the introduced aliens have greatly increased it. Much smaller numbers of native rheas, guanacos, vizcachas, and maras are faced with a population explosion of native pumas and foxes in a newly hazardous ecosystem.[50]

This unnatural situation persists despite the fact that tons of hares and foxes are hunted by humans for their meat and fur, respectively, every year. Fox skins alone represent more than $1 million in annual income for local people. Each ranch worker who hunts can count on catching ten to twenty foxes each year, which, at $10 per skin, represents more than 50 percent of a monthly salary—although cartridges are not cheap.[51] Even though illegal in most places, this hunt has become so important to so many rural people that its full extent is unknowable. An overall annual trade of $200 million is estimated. From 1949 to 1956, at least 54,000 culpeo skins were exported from Argentina, while from 1976 through 1979, 3,500,000 gray fox pelts, 340,000 Geoffrey's cats, and 78,000 pampas cat skins were exported. Despite such figures, which have declined in recent years, some fox and puma hunting, combined with a reduction in the numbers of hares and rabbits, may be necessary for the recovery of native herbivores. Humanity's changes in North American ecosystems have made threats of such improbable species as cowbirds. Over 130,000 parasitic brown-headed cowbirds (*Molothrus ater*) have been dispatched in rare Kirtland warbler nesting areas since 1972, while wolves and pumas are being killed to protect the gravely endangered Vancouver marmot (*Marmota vancouverensis*).

In the United States, where wolves still live or have been reintroduced, they restrict the number of coyotes in a region, and coyotes in turn appear to keep the number of foxes down. In the places where culpeos are numerous, they probably keep gray fox numbers down. It is the gray fox, not the bigger culpeo, that is most likely to prey on small mammals and birds and ground nesters like martinetas. Because the ecosystem will never again be the same as it was, except in very large reserves, humans may have to substitute for some predators to "manage" populations of herbivores to maintain their forage and carnivores to sustain the herbivores. However, the poor people who now do most of the hunting receive only a sliver of its profits. Andres Novaro and Susan Walker believe that institutionalizing fox hunting on a scientific basis and ensuring that its profits are fairly shared may be the most reasonable next step to mollify the sheepmen and provide relief to besieged native wildlife. Andres found that where sheep were suddenly removed, predation on native animals became calamitous.

Joaquín, an elderly man hired to care for Fundación Patagonia Natural's protected estancia, La Esperanza, told us, with a downcast expression, that becoming a conservationist and refraining from the custom of killing almost every sizable native species is an extremely difficult lifestyle change. On other estancias, he had frequently hunted guanacos (266 in one year), choiques, pumas, foxes, hawks, and so on. To his family, which looks forward to his latest bag, he explains, "Only the puma has *asados* [Argentine barbecues] at La Esperanza." He's right. Puma numbers are rising, and they have killed several of the estancia's guanacos.

Although the European hare is already a monumental pest on the steppes, more bunnies are on the way: European rabbits (*Oryctolagus cuniculus*) have also colonized Argentina. Australia's rabbit experience is relevant. In 1859, an Australian named Thomas Austin, of Winchelsea, Victoria, imported twenty-four wild rabbits from England. He released them for sport hunting. Within a few years, they had overrun Austin's property and spread into neighboring districts—they had been loosed in a part of the world where they had no natural predators, no parasites. By 1880, two million acres of Victoria had been eaten bare. The rabbits advanced over the rest of the landscape at the rate of seventy-five miles a year, indiscriminately destroying grasses, bushes, and young trees. Then, in the 1890s, Australia suffered a major drought. That, combined with the rabbits' consumption of grasses, left the ranchers' big sheep herds nothing to eat. In ten years, thirty-five million sheep perished; nearly half the total, sixteen million, died in 1902 alone—but not the rabbits. Finally, myxomatosis, a virus devastating to rabbits but nothing else in Australia, was brought in by ranchers. It was 99.9 percent effective, but 0.1 percent of rabbits proved to be resistant. Rabbits soon reached over three hundred million again. Now, a new hemorrhagic fever–type virus has been introduced by the government to kill them and, again, appears 99.9 percent effective. We will see.

Back in Patagonia, as if what hares, rabbits, and sheep growers have wrought is not challenging enough for native animals, the European pig or boar (*Sus scrofa*) is spreading throughout the more heavily vegetated areas of the steppes. It was introduced on an estancia in the province of La Pampa in 1906 (along with red deer) and is becoming a ruinous ecological force. This is a big, destructive, rooting animal that does not confine its attentions to roots. It eats any small animal it can overpower, takes small tortoises and birds' eggs, alters vegetation, and even destroys flamingo colonies. Andres Novaro tells me that in some areas, its foraging endangers regeneration of the ancient alerce trees, found only in the west of the province of Neuquén in Argentina and in southern Chile. Nevertheless, some ranchers are encouraging the pig for trophy and commercial hunting, native flora and fauna be damned.

Such alien animals are introduced for sport rather than food or skins. The foreign pig, hare, red deer, Asiatic axis deer, Indian black buck, and even trout are considered Argentine sport and trophy species. Almost every one of the introduced species has proven destructive to local fauna and flora. Globally, the economic importance of trophy hunting (except fishing) is small. Even in Hungary, a major supplier of trophy hunting, it is only 0.0005 percent of GNP.[52] Nevertheless, Argentine ranch owners release whatever animal they wish on their estancias, although they know it may spread to the property of their neighbors with harmful consequences. These remarkably self-indulgent decisions are far from being an Argentine idiosyncrasy. The nineteenth and early twentieth centuries were a time of species introductions by Europeans around the world, most famously in New Zealand, Hawaii, and the continental United States. "Acclimatization societies" were organized to foster bringing animals "from home" for sentiment or simply to shoot or fish—in Patagonia, almost entirely the latter. Most such introductions failed, but, of those that didn't, a high percentage have become pests—no natural predators, no parasites.

One of the latest invasive disasters in Argentina comes from mink being released from fur-breeding farms. Turned out south of Bariloche in the 1980s, they are now well down the Chubut River, killing waterfowl as they go. They have also been released in Tierra del Fuego, along with beavers, whose burgeoning population is said to have damaged as much as 15 percent of the forest.

David Pimental and his colleagues at Cornell University calculate that nonindigenous plants and animals in the United States "cause . . . losses totaling approximately $137 billion per year."[53] Of that, feral pigs alone cost landowners and others about $800 million each year, and their effects can be quite sad. Feral pigs in Hawaii are destroying whole forests as well as native wildlife. Nevertheless, exotic-animal introducers argue that new income from exotics outweighs any negative effects. Rarely is that so—and whose income? Any positive effects are usually private, while the huge negative effects end up on the public's tab. In Patagonia, the effect of introducing exotic animals doesn't stop with the steppes' dry land. Even fish can be rabbits, and even parks can be victims.

In 1960, when Bob Goelet and I camped at the exquisite Laguna Blanca park in the foothills of the Andes near Zapala, there was a small flamingo colony and a substantial black-necked swan population, pegged seven years later at 2,108 by Argentine scientist Juan Daciuk.[54] Today, that lovely lake, a national park, has a handsome interpretive center but a diminished flamingo colony and less than a dozen swans, according to a count Graham and I made. The center's signs tell the story: Perch were introduced to the lake in the 1960s, apparently without the park administration's permission. They ate the aquatic plants, the native fishes, and the food of everything else—once again, no natural predators, no parasites. Almost all the birds are now gone, and the park's rangers have no idea how to rid the lake of perch.

Grazing goats, cattle, and horses are also degrading the park's land and the lakeshore waterbird nesting areas, though ways of getting rid of those animals are no mystery. But the introduction of fish is complex, and its impacts are not always predictable. From 25 to 50 percent of freshwater fishes caught in the United States are now from introduced populations. It is clear that fish should rarely be introduced anyplace where there is anything else worth saving, which puts such matters on the problematic slopes of cultural values, understanding, and judgment.

Despite wild animal aliens and more abundant pumas, foxes, and eagles, uncontrolled grazing by cows, sheep, goats, and horses is the most serious conservation problem facing Argentine parks and reserves. Usually, only a few livestock owners are involved, but their eviction is averted by local politicians ignoring their public responsibilities, with the result that some truly marvelous "protected areas" are threatened. For example, near the border of Neuquén and Mendoza, where "giant cones of old volcanoes thrust up into the sky and huge canyons cut deep into a rugged land shaped by the action of incessant winds over the millennia," as Susan Walker once described the region, is a wide, shallow lake called Llancanelo. It has a colony of Chilean flamingos (nine thousand nests in 1990), breeding black-necked swans, coscoroba swans, and many other waterfowl. The area is a Mendoza provincial reserve, but its shores and marshes are infested by masses of goats and cattle grazing in what should be waterbird nesting areas, and the flamingos there are now endangered by European wild boar.

Farther south, almost within sight of Llancanelo, is another reserve, this one immense as well as gorgeous. It covers some 400,000 hectares (988,000 acres) dominated by the 12,073-foot (3,680-meter) Cerro Payunia, for which it is named. The reserve lies in a transition region between altiplano and steppe, and it is embellished with handsome formations of red granite, extraordinary views of the Andes, and nearly white sand deserts adjacent to green savannas. It has at least fourteen thousand guanacos—in the most arrestingly beautiful setting for them in all Argentina. There are colonies of the bizarre and disappearing plains vizcacha (*Lagostomus maximus*), and chunky mountain vizcachas (*Lagidium viscacia*) skip back and forth along the granite cliffs. Pumas are here, and condors are abundant.

This remarkable area is threatened not only by large numbers of grazing cattle, but also by oil and mineral interests and by local attempts to turn more of the area over to private ownership. Payunia is world class in beauty and importance and should be a UNESCO World Heritage site. A particularly glaring example of destructive stock grazing is taking place nearby at the Cerro Tromen Reserve in Neuquén, which holds jewel-like wetlands.

Laguna Tromen is the reserve's main gem, a little wetland at the very foot of Cerro Tromen (4,114 meters, or 13,497 feet). We visited it at the suggestion of a ranger in the town of Chos Malal and found more than 350 pairs of black-necked swans, each with cygnets, pushing their way through the laguna's colorful salad of edible water plants. It was almost like Laguna Blanca when Bob Goelet and I first saw it. There were masses of Argentine ruddy ducks, rosy-bills, red shovellers, speckled teal, brown pintails, silvery and white-eared grebes, three species of coots, and even coscorobas. Flights of waterbirds were taking off, landing in front of us, circling, constantly coming in and flying out. It is the densest, most diversified waterfowl wetland I have ever seen, stunning! Viewing had a special frisson of excitement because not everything in such lakes is known. In 1974, Maurice Rumboll discovered a wholly new and spectacular bird, the hooded grebe, on a little-known mountain-edge laguna. A few years later, Graham, Kix, and I found the lake and saw the grebe. (We learned that the laguna's name, Los Escarchados (The Frozen Ones) was well earned—several people had frozen to death there. Laguna Tromen wasn't just cold; it was like the Dakotas in January with a 50-mph wind.)

This exquisite little wetland is being destroyed by the privately owned cattle, horses, sheep, and goats. The horses and cows wade into the nearby marshes and eat the fragile vegetation that protects the few areas birds have to nest, munching reeds at one end and defecating great lumps and pies into the water at the other. Meanwhile, myriad goats and sheep work the shores, chewing off any vegetation they can reach. It is tempting to justify sanctuary degradation by pointing out that some of the people degrading it are poor. But they could be helped in other ways, not by destroying irreplaceable natural resources. It is just as possible that a corrupt official has been paid off, or simply doesn't care. With Llancanelo and Payunia, and perhaps nearby Auca Mahuida someday, this reserve forms a mosaic of wildlands that could be a marvelous source of pride and enjoyment for generations to come, but all are being degraded. Grazers are taking private advantage of public property and debasing it over much of Argentina.

Although it is clear that the future of wild animals will depend, more and more, on parks and their active care, few park authorities anywhere have much experience actually managing wildlife populations. Argentina is no exception. Until recently, it seemed enough to simply protect wild animals and reserves from excessive exploitation, but today's diminishing wildlife populations and constricted reserves require focused stewardship. The interactions between the native species we should be trying to sustain; the impact of invasive species, native as well as exotic; and the management of predators have created a convoluted list of needs. The role of top predators is particularly complex.

In the United States' Yellowstone Park, pumas kill an elk about every nine days throughout the year, which makes some wish to control the pumas. However, a superabundance of herbivores, without the top predators, might overeat the vegetation, interfering with its reproduction, while middle-sized predators multiply, exterminating small creatures of all kinds. In fact, an absence of top predators in wildlands "appears to lead inexorably to ecosystem simplification accompanied by a rush of extinctions."[55] In closed-park situations, human beings must often become the predators. For example, South Africa's 19,943-square-kilometer Kruger Park can support no more than 7,500 elephants, which are superherbivores, without gross habitat destruction and, thereby, the extinction of other wildlife and eventual diminution of the elephants themselves. To keep the population within supportable bounds, the park has conducted an elephant control program for decades. However, "animal rights" groups stopped the control program with protests, resulting in a current elephant population of 12,000. A similar situation has resulted in the demolition of Kenya's once magnificent Amboseli Park. Its forests are gone, along with their wonderfully diverse wildlife—hornbills and hoopoes, gerenuks, giraffes, and much much more. In limited spaces, even the most desirable species will probably have to be controlled. In the Southern Cone, that could apply to guanacos and, certainly, to some native predators, be they pumas or eagle buzzards. Native herring gulls (*Larus argentatus*) and black-backed gulls (*Larus marinus*) have become predators of terns in the United States, while superabundant kelp gulls (*Larus dominicanus*) threaten cormorants, terns, and even whales in Argentina. The principle is the same whether applied to pumas, gulls, or hippos—understanding carrying capacity and interspecies effects requires knowledgeable and sometimes distasteful management tailored to each situation to sustain biodiversity.

When we Wheel People replaced the Foot Indians, we changed the rules. We took over the habitat itself, massively reducing its wildlife-carrying capacity, the number of staterooms in nature's boat, and then brought in vast numbers of incompatible new passengers and changed the character of the accommodations. Today, on the Patagonian steppes, native carnivores are thriving while many native herbivores and omnivores are in steep decline.[56] Ecologically, humans have long been the steppes' top predator, and I suspect we must continue that role at some level if only to protect the native species—but we will have to proceed very knowledgeably and very transparently.

West of the steppes, where the condor soars high among volcanos, where high-altitude vicuñas replace guanacos, and where Mr. Darwin's rhea has changed sufficiently to be called suri, the steppelands grade into the unearthly world of the altiplano. It is a little-known region with a special set of human-wildlife relationships and a place where the landscape species are not big mammals but improbably beautiful flamingos in the most dramatic desert landscapes in the world.

*It is said that the Phoenicians bartered dried flamingos to the barbarians of Cornwall in exchange for tin, and to the people of the Netherlands for amber, passing the skins off as those of the fabled Phoenix and thus an article that might transfer immortality to its owner. Man has always found something magical about flamingos.*

Sir Peter Scott

---

# 7
# *The Altiplano and the Flamingos*

The dry, grimy city of Antofagasta squats ungracefully at the edge of the Pacific in northern Chile, near the northwest corner of the Southern Cone. Abandoned villages punctuate the blazingly bright dusty roads leading to the decrepit municipality, while eroded ruins of deserted huts and stores and sand-drifted graveyards interrupt immense vistas swept bare of vegetation by the cutting winds of the coastal desert. Much of the town appears to be a poverty-stricken slum of waterless hovels, further degraded from the time of my first visit in 1960, twenty-eight years before. No surprise.

The city was built in the parched Sechura-Atacama Desert, one of the driest in the world. Rain here is practically unknown. The desert runs along the entire southern coast of Peru and down most of northern Chile, almost two thousand miles from end to end. Originally a fishing village, Antofagasta expanded dramatically in 1860 with the discovery of silver. Now the silver is mostly gone, but more than

250,000 people stay on, clinging to a faltering economy that once again is based mostly on fishing but also on nearby copper and nitrate mining. The town's potable water resources, a few wells and a branch of the Rio Loa, are utilized to the fullest, but many households are lucky to get fresh water three times a week. The seriousness of the problem is clear from the fact that Antofagasta is considering seawater conversion, water recovery from fog, and reuse of sewer water. More people live in Antofagasta than the environment can sustain, but restricting growth is an action that seems beyond human imagining. The names of nearby "lakes" tell the story: Lagunas Secas (Dry Lakes) and Salar Mar Muerto (Dead Salt Sea). Globally, fresh water amounts to less than one-tenth of 1 percent of all the earth's water, and it is rarely where we want it.

There is nothing unusual about this dilemma. All over the world, poverty-stricken people are living in inappropriate places in insupportable numbers, becoming increasingly desperate in an effort to survive where they have put down roots. Over the last seventy years, the world's population has tripled, and water demand has increased sixfold, that alone threatening wildlife far beyond city limits. Yet not far from desiccated Antofagasta, there are perfectly adapted waterbirds, including three extraordinarily beautiful but little-known flamingos.

Most of these birds are found in the cordillera, far above the tree line in the bone-dry puna and altiplano, mostly above 11,000 feet (3,350 meters), and in the Andes plains between its peaks. They live on a few soda lakes, or salars, broad, shallow, desert basins fed by small streams with no outlets. Incoming waters simply evaporate in the sun. The feeder streams, originating from mountain snows, glaciers, and undependable mountain rains, are laden with minerals captured during their tortuous descent to the salars. Wherever they settle and evaporate, they deposit dissolved burdens of glaring white salts and form dry "pans." Where they don't evaporate, they create salars with a briny soup of corrosive nitrate, carbonate, lithium, sulfate, and manganese, a caustic mix in which specialized diatoms (siliceous unicellular algae), nematodes (tiny threadlike worms), and other invertebrates proliferate and become food for flamingos.

Flamingos are everyplace and noplace in the Southern Cone, from the highest mountain lakes to desert seashores and, once in a while, all the way out to the Falkland/Malvinas. They feed near shepherds, sheep, maras, and guanacos on the lagoons of the steppes, and many nest there, alongside gauchos and cows in the lakes of the pampas, and next to vicuñas, horned coots, and llama herders in the Andes. They thrive in waters parboiled in the hot magma of volcanic cones, leached through salt-laden soils, and frozen nightly in frigid winds at salars, salinas, and salitrales. They can fly great distances in savage winds, yet, for them, the usable part of the world is tiny. They are tied to isolation and to colonial breeding, like so many other creatures in the Southern Cone, and to those island-like basins of saline water in the barren steppes and the dry altiplano, and that is where they build colonies and rear their chicks. In their own way, they are as tough as penguins and tougher than tortoises.

The majority are "Chilean" flamingos (*Phoenicopterus chilensis*), but high in the mountainous altiplano between the volcanos, the comparatively abundant (perhaps 100,000) *chilensis* is joined by the two rarest and most beautiful flamingos in the world, the Andean flamingo (*P. andinus*) and James' flamingo (*P. jamesi*). The Chilean flamingo is pink and white, with scarlet red wings and slate blue legs. James' flamingo is much smaller, with dark red legs and face, a finer bright yellow beak, and a capelike mantle of rose-red plumes over an underplumage of white—an exquisite bird and very much a high-altitude one. I found that its wing area was even larger than that of the much bigger greater flamingo of Africa's Rift Valley—thin air, broad wings. The Andean flamingo, largest of the three that breed in South America (the Caribbean flamingo only visits the shores of Colombia, Venezuela, and Surinam), has bright yellow legs, a rose-red neck tinged with magenta, which grows darker on the breast, and prominent black tertial feathers, which create a black triangle near its tail. Its face is pale yellow with a curious purple spot just in front of its eye.

Flamingos are ancient, much older than the Andes and the steppes. Their line goes back nearly forty million years, almost as far as that of rheas. They seem to have evolved in the Old World, where the greater and lesser flamingos dwell today, and come somehow to the New World only a few million years ago. Both James' and Andean flamingos live primarily between the east and west edges of the Andean cordilleras, mostly above 7,000 feet (2,100 meters)—perhaps mostly above 11,000 feet (3,300 meters). And, of course, humans kill them.

They are killed for food, for Indian medicines, and for their feathers as decorations, and their eggs are stolen. In the 1960s, when I became familiar with this situation while capturing several birds for the zoo, I worked to have Bolivia's Laguna Colorada made a reserve. Later, WCS sponsored a global flamingo census by U.S. ornithologist Philip Kahl. Then Felipe Lariviere, Argentine national parks president at the time, asked me to help in his program to create reserves. Our efforts nearly got us killed among the Andean peaks in the most frightening small plane flight I have ever made, but they resulted in the creation of Lago Pozuelos National Monument. This spectacular lake, 11,500 feet (3,500 meters) up in the far northern Argentine province of Jujuy, is an exceptionally important feeding area for all three of South America's flamingos. Until recently, however, the most extensive efforts to preserve the flamingos took place in Chile.

In 1985, the Chilean Forestry Department (Corporación Nacional Forestal, or CONAF) appointed biologist Mario Parada to explore flamingo conservation, and to help in the endeavor, WCS contributed to the cost of Parada and several rangers, a trailer for rangers to live in, watchtowers, trucks, and other essential equipment. In three years, Parada and his rangers got the first extensive data on the birds' habits and whereabouts and stopped the poaching at critically fragile flamingo colonies. Eventually, Chile created a far-flung national flamingo reserve system, but we soon realized that the birds move freely between the highlands of southern Peru, Chile, Argentina, and, especially, Bolivia. No one nation could possibly preserve them by itself. However, Parada had begun to build that foundation of fact without which lasting conservation is always elusive.

Graham, Kix, and I traveled to Chile in December of 1988 to review the flamingo project's progress. We met our hosts, and then, guided by Alfonso Glade, slender, black-mustachioed, irrepressibly good humored, and impressively well informed, we headed north to Antofagasta, where we met Mario Parada and Juan Pablo Reyes, CONAF's regional director. Mario, the quintessential puna biologist, is darkly intense and a fund of flamingo facts. Juan Pablo, his boss, is a big, handsome man, prematurely gray, with a world outlook and a sly but gentle sense of humor.

We headed for the venerable village of San Pedro de Atacama, driving through the spectacular Passo Barros Arana (11,154 feet, or 3,400 meters) and past the Valley of the Moon. "Moonscape" is a fair if overused description of the topography approaching San Pedro. Distant objects are lost in arid mirage, and breath dries in the throat. Except for rare tufts of grass, it is plantless.

San Pedro proved to be postcard picturesque, dusty, and poor (population 911; altitude 7,992 feet, or 2,436 meters, a sign said). It is built immediately below the spectacular black cone of Volcán Licancabur, which, at 19,425 feet (5,921 meters), is a shade higher than Tanzania's Kilimanjaro. The village was settled at least by the 1530s: House numero 11, Casa de Pedro Valdivia, built in June 1540, still stands; Soldier Valdivia marched all the way from Cuzco (Peru) over 450 years ago with a small group of soldiers and Peruvian bearers to find a place to settle. The village church, its high ceiling supported by huge cactus beams, is dated 1557. The town is further distinguished by a little museum founded by the late Father Gustavo Le Paige de Walque, S.J., that houses remarkably preserved mummies of the Tiwanuku-Atacameños Indians, who inhabited this dry, forbidding area seventeen hundred to a thousand years ago but were eventually conquered by the Incas.

By 8:00 p.m., the air was cool, and deep shadows darkened the village around us. The setting sun gleamed back from the upper half of Licancabur, and it was easy to imagine a glitter of Inca gold. Our little party balanced Coke bottles and deadly Pisco sours on a rickety table behind the local hosteria, gazed at Licancabur, and told tales of wondrous birds, mummies, and altitude sickness—and the improbabilities of Chile-U.S. government relations. The bottle balancing reminded Juan Pablo of a Santiago radio commentator's latest bon mot: "The reason I hate America is that the tables don't wobble."

# ACT III

Early the next morning, we picked up a fifty-gallon drum of fresh water for the rangers guarding the flamingos and drove south into the desolate Salar de Atacama, which is at about 7,000 feet (2,134 meters). This huge, shimmering salt pan must be about as close to hell as can be reached by four-wheel drive. By day, its temperature commonly exceeds 120°F; by night, it is cold and windy. A glaring, plantless, slimy lakebed of 1,544 square miles, it appears in places to be made of broken glass. The painfully bright salty mud is relieved by less than 6 square miles of more or less open watery brine, where, clustered in groups of close-set mud nests, is what Mario believes to be the largest remaining colony of the magnificent Andean flamingo, the "Parina Grande."

To a mammalian predator, the wide, waterless salar with its hot, sharp salt and bubbling gases is a formidable barrier. Imagine miles of blade-like salt blocks, black mud, and brine. There is no prospect of shade, or of easing thirst. It is an area of pitiless sun and wind and, until recently, it protected its flamingos from Andean foxes, Geoffrey's cats, and human beings. Now mining of the pan for its lithium, nitrates, potassium, boron, and molybdenum has resulted in the construction of oil exploration roads that crisscross the area. Those roads, bashed through the salar with heavy earthmovers, have given entry to foxes and encouraged local Indians from the villages of Toconao and Peine to ride in with bicycles or walk in with their burros to raid the flamingo colonies for eggs. The raiding had become so extensive that chick production had virtually halted when Mario's project began.

Just as serious a threat to the colony, and potentially terminal, may be a precipitous reduction in the salar's pitiful sources of water. Its major tributary has already been reduced by irrigation projects at a marginal village above it, and the life-giving dribble may be totally preempted for the mining processes of a nitrate company, and even for the city of Antofagasta itself. To make matters worse, a luxury hotel has been built near San Pedro, placing well-funded new demands on the disappearing water.

We entered the burning brightness of the salar near a WCS-sponsored watchtower, treading gingerly—not only to save our boots from salt cuts but also to avoid sinking suddenly through its top layers into the sulphurous gray-black ooze beneath. Incongruously, barn swallows hawked for insects over our heads; I wondered if any were from home. Quietly, Alfonso pointed upward, and we stopped in speechless admiration as a single pair of shockingly brilliant Andean flamingos descended from the east's volcanic heights, high in the deep blue sky, and floated down to the salar—as lightly and brightly as burning ashes from the volcanos themselves.

From atop the watchtower, we could see Andean flamingos on all sides. They fed in small groups, filtering the bitter brine for food and behaving quietly for a time, then raucously challenging one another, chest to chest, in disputes over matters of personal distance and dominance. Courtship and nesting had begun. Elegant marches were in progress—close-packed groups, striding synchronously through the shallow brine, heads held stiff and high. The flocks paraded, pivoted, gave open-wing salutes, and, acting in unison, erected their richly hued neck plumage. The results were sudden bursts of intense color, as though huge magenta flowers were suddenly opening, then closing. Some birds were already sitting on nesting mounds in the shallow water. The gray mud platforms, several inches high, protect their eggs (one to a nest) from water level rises and were concentrated in groups. Mario said incubation requires about thirty days, and chicks are big enough to fly in April. As always when near flamingos, their constant chatter struck me. A rasping "heh-teh-teh, heh-teh-teh" was the Andean flamingo chorus; a resonant "to-co-co, to-co-co" came from a single pair of Chilean flamingos nearby.

Limnologist Stuart Hurlbert, of Humboldt College, studied the watery "grazing" of *andinus* and found that experimental exclusion of the birds from a portion of one lake resulted in amazing increases (one hundred times) in the large diatoms (*Surirella* spp.) and nematodes in the sediments they eat. The birds can overgraze a lake just as sheep or cattle can overgraze a pasture. But the three South American flamingo species differ from one another in their feeding apparatus. The apparent "teeth" on the comblike external lamellae of the Chilean flamingo's beak strainers are spaced between thirteen and sixteen to the inch on those I have measured, and those flamingos feed largely upon small invertebrates, especially brine shrimp. Andean flamingo teeth number about twenty-three to the inch, while those of the James' flamingo are very fine, about fifty-four to the inch in the three birds I have checked—diatom eaters. This beak comb is backed up by a complex grid of tiny keratinous serrations in the mouth, which trap the bird's minute foods, serving flamingos much as whalebone serves whales. A stream of water "cuisine" is pulled through the beak and expelled from this strainer by the pumping of the bird's plunger-like tongue, which has yet another comb to pull the catch from the inner-beak serrations into the bird's esophagus. Different-sized filters allow the three species to feed more or less noncompetitively side by side, extracting different foods, as has been noted for Africa's greater and lesser flamingos.

Chief ranger Ernesto Cerda Candia, a tall, impeccably uniformed man with one of those finely chiseled Chilean visages that sticks in the memory, indicated a line of sight between the slopes of Volcán Lascar, venting black smoke above us, and a peak to the west. As I teetered on a lump of salty mud below the volcanos, Ernesto explained that I was precisely astride the Tropic of Capricorn and pointed south to a point that egg poachers must pass to gain access to the birds in the salar. There, CONAF has a temporary ranger station.

Fifty minutes of bouncing slowly over mining truck tracks brought us in sight of a disembodied aluminum blob dancing in a heat mirage. Closer, the blob became a small aluminum trailer emblazoned with the words "Donación de Sociedad Zoologica Nueva York," which housed two rangers nattily attired in crisp white shirts with NYZS decals (the New York Zoological Society became WCS in 1993). These young men, trained by Parada, came from a nearby Indian village to work for CONAF seasonally, and they have been very effective in stopping egg poaching—between eight thousand and thirteen thousand birds nested in the salar that year.

On our next trip, in 1996, we came face to face with yet another of the flamingos' life-threatening problems. Jumping out from behind a salt block, a fox suddenly splashed its way into the middle of the nesting birds. It flushed the entire colony as it went from one nest mound to another seeking eggs and chicks. It clearly had done this before and knew where to go, but this time it was too early—the birds had not yet laid. But the damage was done; the flamingos abandoned the colony for the entire year. Andean flamingos are extremely sensitive to disturbance.

By 1996, general agreement had been reached on the need for a four-nation "flamingo summit." There was growing evidence that the James' and Andean flamingos were in trouble. And despite Chile's protection program, Mario Parada estimated the maximum summer population at about 75,000, only a portion of which nest there. The majority of the birds could be lost elsewhere. Besides, I had begun to see these two extraordinary birds with new eyes, as flagship or landscape species. Their populations are an indicator of the laguna ecosystem's health and productivity, and the remarkable efficiency with which they harvest laguna invertebrates affects the lake's ecology. Their protection could be key; it could gain attention for the need to protect the overall altiplano lake ecosystem.

I asked Peruvian biologist Marianna Valqui-Munn to try to pull together the summit that same year—a tough assignment. Such a collaboration is not a natural one, for Bolivia and Chile often dote on past grievances and do not have diplomatic relations; plus, relations between Argentina and Chile are always in flux. However, she arranged for the meeting to take place in San Pedro de Atacama and have expert delegations from all four countries—Peru, Argentina, Bolivia, and Chile.

The summit was a success. Attendees assembled information and ideas and cemented international friendships. The meetings produced an awareness of the need for collaborative conservation of the birds and their salars. They laid a foundation for the establishment the following year of Grupo para la Conservación de Flamencos Altoandinos (GCFA; High Andes Flamingo Conservation Group), for which WCS and its flamingo expert, Felicity Arengo, have been an organizing force.

Each of the scientists who attended the meeting learned to distinguish the look-alike juveniles of the three species, all of which are gray with black legs, through direct field instruction. Well before first light, the workshop's 4x4s ground slowly up Licancabur's southwest flank toward the Bolivian border and the high puna. Ten strong, we were on the way to the main nesting grounds and rediscovery site of the James' flamingo, which was long thought extinct until found again in the 1950s. Our destination was a magnificent Bolivian salt lake of bright red waters covering sixty square kilometers (twenty-three square miles), Laguna Colorada.

We crossed the ramparts of the Cordillera Occidental through Passo Purico (15,748 feet, or 4,800 meters), directly south of Licancabur where the track curves east and then north to Laguna Blanca and a Bolivian border post, a bedraggled procession of crude sheds perhaps 250 feet long. From the post and the white lake, where Andean flamingos were feeding, the track twists past jade green Laguna Verde, where I had watched rare horned coots (*Fulica cornuta*) many years before and feared I would have to walk out. There was no border post then, and when we wished to leave, our venerable truck, a modified WWII army ambulance, would not start. We did have a crank, though, and for two days we cranked through sleet, biting cold, and howling wind and finally got under way. Our new 4x4s had no cranks, but they didn't stop, either—except when we broke a spring. From Laguna Verde, the track went through the bleak Kharmapampa, past the Salar de Chalviri, and through the volcanically active plateau of Sol de Mañana (15,760 feet, or 4,804 meters).

The altiplano is intimidating, not in miles or kilometers but in altitude and the paucity of ways for plants and animals to make a living. Although much of it lies north of the Tropic of Capricorn, it is the antithesis of the lush rain forest in the lowlands to the east. The Kharmapampa's substrate consists of stony shingle, a coarse volcanic sputum interspersed with splinters of obsidian. Our route traveled between truck-sized boulders of wind-polished black basalt, brick red dunes, and treacherously soft sandpits, passed towering volcanos, vast treeless plains, strange buttes and spires. Vegetation was nearly absent, and the trucks looked like ants in the desolate grandeur of the vast ochre plains and black and red volcanos.

It was thirty-six years since I had been to Laguna Colorada. At the urging of famed ornithologist Roger Tory Peterson, I came here to catch and bring back to the Bronx Zoo the rediscovered James' flamingo.[57] Roger had photographed it two years before, becoming the first person to have seen all six species of the world's flamingos. (Ornithologist James Fisher, a fiendish friend, subsequently introduced Roger at an international congress in 1958 by extolling his distinction as the only human to have seen all six flamingos—and concluded by saying, "Of course, when you've seen one flamingo, you've seen them all!") Roger aimed me at Laguna Colorada and gave me his Laguna-proven waders. So few people had visited the red lake in those days that my guide and colleague, Chilean entomologist Luis Peña, was able to point out the two-year-old tracks of Roger's truck.

The first James' flamingo known to science was shot near Isluga Volcano in Tarapaca, Chile, in 1850, but it was recognized as a new species only after British businessman H. Berkeley James financed an expedition that obtained more specimens in 1886. After that, nothing more was heard of the bird. In 1956, Robert Allen wrote, "No actual breeding sites, past or present, are known . . . we cannot but wonder if James' flamingo still survives."[58] The seven-decade hiatus was a challenge that motivated A. W. Johnson, his son Bryan, Dr. Francisco and Erika Behn, and W. R. Millie to make four expeditions and finally find the James' flamingos at 4,278 meters (14,035 feet) in Laguna Colorada in 1957.

The average temperature in the altiplano is only 35°F, and the daily variation is from 5°F to 68°F; but the altitude is the most daunting and sometimes sickening condition for those with no time to become acclimated. The graceful vicuñas, which raced off so effortlessly upon our approach, average about fourteen million red blood corpuscles per cubic milliliter versus our five or six million. That explains much. The pressure of inspired oxygen at 14,000+ feet is only about 56 percent of its value at sea level. Even if you are well adapted, oxygen is harder to get.

Eventually, we paused to disassemble our trucks' air intakes, remove their "puna" screens, and pour out accumulated dust and sand. Finally, we drove over the last pass and into Laguna Colorada's immense mountain-walled basin. The lagoon itself appeared as a huge horizontal sheet of scarlet, its edges bordered in shining white, which proved to be salt. Glacier-like fingers of icy sodium sulfate projected from its shores, and salt devils whirled at its center; but distant motes of pale pink in the blood-red waters eventually resolved themselves into flamingos, thousands of flamingos. Black cliffs bordered part of one shore, high shingle and sand banks bordered another, and grass-green bogs extending from warm volcanic springs edged a third. At 14,035 feet, the waters are only 75 feet lower than the top of the United States' Pike's Peak and almost 400 feet higher than Switzerland's Jungfrau.

Dropping our supplies at the ranger station, new since my 1960 visit, we began a survey. On my previous visit, I found the shores littered with obsidian arrowheads. People have been hunting here for a long time, and Andean geese, Andean avocets, goose-sized giant coots, puna and sharp-winged teal, and masses of shorebirds make it a migratory rest stop. Vicuñas come to drink from the springs and graze on lakeside vegetation, and Andean foxes and wild felids are attracted by everything else. Laguna Colorada is the most magnificent representative of what all who know the altiplano hope to save there.

The Quechua-speaking and Colla peoples who live in this part of the Andes dwell lower than Laguna Colorada and are dependent on their llama flocks for much of their sustenance in a land largely unsuitable for crops. The 2,762-square-mile Reserva Nacional de Fauna Andina Eduardo Avaroa that Bolivia has set up to protect Laguna Colorada and nearby lakes has only six hundred inhabitants.[59] But mining interests and plans to produce geothermal power are threatening the wildlife anew, even as authorities seek to protect it from local hunting and egg collecting. More complicated is the arrival of adventure tourism. Over three thousand tourists managed to visit the lake in 1996. No one could possibly distinguish anyone's track now.

Mindful of the altitude, I had taken the precaution of bringing a few pages from Wilkerson's *Medicine for Mountaineering* and passed them on to my colleague, veterinarian Billy Karesh. That night, in the ranger's station, Billy read the list of altitude sickness symptoms to our workshop team: headaches, dizziness, shortness of breath, nausea, disturbed sleep, anxiety attacks, Cheyne-Stokes breathing, life-threatening edema, and more. Soon, the erratic blowing and inhaling called Cheyne-Stokes respiration interrupted the bunkroom murmuring. Headaches were noticed, and stomachs became queasy—at least until Billy shut up. Fortunately, I was in another room and slept well.

For as long as we have known the birds, the main James' flamingo breeding area has been Laguna Colorada, where I counted between six thousand and eight thousand in 1960. But, in 1972 and 1973, when Phil Kahl made his flamingo survey, there were none there at all. Our workshop team estimated nearly thirty-five thousand during this visit, but the total for all its populations remained unclear. Bolivian expert Omar Rocha reported that of nine thousand chicks hatched in 1993, only two thousand fledged. I counted the remains of thirty-five chicks in fifty paces along the shore—none with signs of predation. If disease is a contributing factor, clues had not been collected. I was troubled to discover that the family of one of the Laguna's rangers was keeping chickens, almost certainly a reservoir of avian disease.

A race seems to be on to find some way to exploit Laguna Colorada other than tourism—a new geothermal facility appears to have been sited nearby with a view toward doing the greatest amount of damage to the breathtaking beauty of the lake. During our visit, a Mexican engineer was completing an evaluation of this elaborate project for the Bolivian government. "In a word," he said, "the project is nuts." The lake is not too high for ecotourism by the hardy, if they can be controlled. However, the lake's two rangers told us of growing problems with inconsiderate tourists, who frightened the birds and deliberately went where they were prohibited. Yet ecotourism may be the saving of the altiplano lakes. Pride in reaching such a difficult destination, as well as its unusual beauty, can be a basis for the essential public constituency. At the same time efforts to protect the flamingos are growing, however, Chipayan Indians at Lago Poopo are catching them for feathers and to eat. Omar Rocha believes that the major breeding population there is now collapsing.

What needs to be done to ensure that the volcano flamingos and their surreal wildlife-filled lakes have a reasonable chance of surviving the next twenty-five years? The next two hundred years? Can poached-out salars be recolonized? Can water regimens be managed to keep the lakes from drying out? Can preservation from mining, egging, and geothermal schemes be won? The answers depend not only on developing an effective constituency but also on clearly identifying the lakes most critical to the birds.

Flamingos are not migratory; they are nomadic. Felicity Arengo says they are very strong flyers and on the move much of the time, shifting between and within a mosaic of rare, scattered lakes whose food resources are constantly changing. In collaboration with Sandra Caziani of the Universidad Nacional de Salta, Felicity placed satellite radios on several at Salar de Atacama and Lago Pozuelos and learned that the birds think nothing of distance, covering over seven hundred kilometers between Bolivian and Chilean lakes in four days or less.

At Chile's Salar de Tara, where about six thousand James' were nesting when I visited in 1988, the waters are rumored to be in danger. The lake is fed by a small branch of the Rio Zapaleri, which arises in Bolivia, grows in Argentina, and ends at Tara in Chile, but even there is no more than a stream. Nevertheless, Antofagasta, squatting 205 miles away, is eyeing that anemic trickle to help alleviate its chronic water shortage. We were surprised to learn that deep underneath the Bolivian department of Potosi and the town of Uyuni, home of some of the best flamingo lakes, are aquifers with a lot of water. We were even more surprised to learn, in 2003, that Bolivian politicians were planning with mining officials to use that water, but not for the local people in that dry and dirt-poor area. Their idea was to allow mining concessionaires to export the water to Chile rather than use it to help local farmers. No studies have been conducted to determine whether the aquifers would ever replenish themselves, but it seems unlikely. There is also a great deal of recently discovered natural gas in the Bolivian Andes, and that, too, was scheduled for export—the distribution of its benefits unclear. For the moment, the Indians have put a stop to both plans.

Many nesting salars hover at the edge of total evaporation—or flooding. In the first instance, their value for wildlife simply disappears. Because of global warming, experts predict that all tropical glaciers, those between the Tropics of Cancer and Capricorn, which provide much of the flamingo lake water supply as well as that of various villages, will have vanished in twenty years. Nor should the coastal towns of Argentine Patagonia ignore the implications of climate change. Virtually all of their drinking water is derived from rivers fed by the previous year's Andean snows and by glacial melt. In 1980, when Felipe Lariviere led that Argentine parks mission to Lago Pozuelos (155 miles northeast of San Pedro de Atacama on the Argentine side of the mountains), we found the lake's waters high and filled with thousands of flamingos. On my next visit, in 1981, Pozuelos was nearly dry, with only 4,500 Chilean flamingos, some groups of Andeans, two James', and virtually no young of the year in the remnant pools at its center. But it is not unusual for flamingos to miss several years between successful nestings. They seem designed for irregular reproduction.

If you customarily lay only one egg a season and get a chance to nest only once every three or four or seven seasons, you had better be long-lived to raise enough young to at least replace yourself. Flamingos, like people, tortoises, elephants, vultures, parrots, and orange roughy fishes, are long-lived. Some have lived more than fifty years in zoos. That explains the persistence of their colonies in the face of repeated egg poaching. It may also explain how they know where they are going when they make one of their sudden group moves: experience. All in all, the ability to survive on diatoms and nematodes filtered from poisonously salty water (enlarged supranasal glands supplement the salt excretory functions of the flamingo's kidneys), to withstand intense cold, and to fly long distances between feeding or nesting areas makes for a creature singularly well prepared to keep its secrets.

However, if we are going to conserve flamingos, we must learn more. We know little about the health of altiplano lakes. The brews of diatoms, bacteria, nematodes, crustaceans, and tiny mollusks that flamingos eat clearly grow best in strong soup seasoned generously with salt. The two principle ingredients of Laguna Colorada's waters, tested by the intrepid Chilean expedition of 1957, were sodium chloride and sodium sulfate. Although more detailed studies have been done, they have not added greatly to the arsenal of conservation tools. We still have no idea of the main causes of adult mortality or how often the average breeder gets to breed in his or her lifetime.

There is no doubt that flamingos have powerful attachments to their traditional nesting sites, but they seem even more attached to each other, bonded in some way to their hundreds or thousands of companions. Could there be bonds of relatedness in these long-lived creatures, clans of mothers and fathers, sons and daughters, and cousins and uncles—or just familiar faces? As it is for many colonial creatures, social facilitation is an important part of their reproductive cycle, a way of achieving synchronization. Virtually all the flamingos of a colony will lay and hatch their eggs within a few days of one another.

"Very unlikely" is a fair description of much flamingo behavior. Once during lunch in Cordoba, the late Charles Olrog, Argentina's most distinguished ornithologist, described seeing an entire flock of downy Andean flamingo chicks, accompanied by adults, walking down a dry mountain streambed far from any nesting lake. The explanation was probably provided in a similar and even more remarkable event that took place in Namibia with the 1970–71 nesting of Africa's lesser flamingo (*Phoenicopterus minor*) at the Etosha Pan and, luckily, was photographed by Hugh Berry.[60] Thousands of chicks hatched just as Etosha's water began to dry up, but, instead of dying on the spot, they formed a massive throng and marched like a procession of little black-legged powder puffs over the barren pan's 120°F mud flats—for a full month and for fifty miles—to water. Every day, all day long, flight after flight of adults flew to food sources as far as one hundred kilometers away and back again to feed their young. And the chicks were saved.

Late at night in San Pedro, back from Laguna Colorada, we sat tiredly around a large (wobbly) table at the hosteria, restoring our fluid balances with as much cold beer as could be found and worrying about flamingo numbers. In the early 1970s, Phil Kahl estimated that there were about 150,000 Andean flamingos, but given the nearly insurmountable task of an accurate one-man census in the high Andes, he speculated that the species might not be that abundant.[61]

Nevertheless, "How many South American flamingos of which species still live?" was the four-nation workshop's first question. At the end of 2000, the international flamingo group estimated sixty-four thousand for the James' flamingo, and more counts and studies are under way. For the Andean flamingo, the GCFA, which had surveyed 220 high Andean lakes by 2000, had found a maximum population of about thirty-four thousand. That is 30 times the estimated population of the grizzly bear in the contiguous United States, 53 times that of the mountain gorilla in Africa, and 205 times that of the California condor. It is a species that has been known to produce fifteen thousand chicks in a single year at one colony. How can a species with thousands of individuals be understood to be rare by a public constituency, which must help protect them? It comes down to the basic problem facing all intensely colonial species in the Southern Cone: *The number of effectively breeding colonies is almost as important as the number of individuals.* And there are few colonies.

Today there are between three million and five million flamingos in the world, mostly lesser flamingos in Africa. Their collective biomass is over ten thousand tons, about twenty-five hundred elephants' worth. So there are a lot of them, and they eat things nothing else wants. But they can breed only in very special, very remote lakes and lagoons. Ninety percent of all lesser flamingos breed at one Tanzanian lake. Eighty-five percent of all James' flamingos breed at one Bolivian lake. If one pinpoints James' and Andean flamingo breeding areas on a map and draws a line around the dots, the resulting region is a flamingo crescent, a new moon, only about 1,000 miles long on the curve and nowhere wider than 180 miles. It extends from roughly 15°S in southern Peru, through western Bolivia to about 28°S in Chile and Argentina. Its area is roughly 120,000 square miles, about the size of New Mexico. But only a tiny fraction of that mountainous crescent consists of lakes where the birds can feed and breed. The rest is high-altitude desert.

If the flamingos can be protected, a unique and breathtakingly beautiful ecosystem and the wildlife community dependent on it will be preserved, as well. However, for the altiplano lagunas, climate change may prove terminal. The last ten years (1994–2003) of summer temperatures in Europe, where we have good data, were the hottest in the last five hundred years. Two-thirds of the glaciers to be seen in North America's Glacier National Park in 1850 are already gone. Mountain dwellers from delicate flowers in the Swiss Alps to pygmy opossums in Australia and pikas in North America are in trouble—and they have no options.

Politically, sociologically, the altiplano is very different from the Patagonian steppelands, despite some of the plants and animals they share. Because of overhunting and introduced alien species, most of the larger animals on the steppes—guanacos, choiques, maras, and the very rare huemul deer—whose historical impact helped shape the nature of the steppe's plants and each other, have become ecologically extinct there. ("Ecologically extinct" means no longer abundant enough to function in or effect the ecology.) The altiplano, on the other hand, is so inhospitable, so plantless, that no big native grazers other than vicuñas can be found in the dry southern areas joining Patagonia. Almost all wildlife is concentrated in and around the scattered salty lagunas with their exquisite flamingos. Those flamingos are the conservation keys, as previously discussed, more so than vicuñas, which range north into other areas. But the social and political options are far less clear on the altiplano than on the Argentine steppes.

No one nation owns the southern altiplano. Besides being divided among Argentina, Bolivia, Chile, and Peru, it is further separated between various indigenous groups, once largely under the harsh reign of the Incas and then under the worse rule of the Spaniards. Despite the unpleasantness of its climate, inimical to conventional farming, some of its original peoples are still there. They were not killed off like the Tehuelches and their kin. Much land, mostly considered "worthless," is owned by governments, and such land tenure as is claimed by cultural groups is not commonly agreed upon. What is common among the altiplano's thin covering of indigenous peoples is extreme poverty.

One result of the poverty and land tenure issues in Bolivia has been the emergence of newly assertive Indian movements, which contributed to the 2003 overthrow of Bolivia's president Gonzalo Sánchez de Lozada. The Indians' new political awareness is clearly aimed at securing equal rights for those long-discriminated-against populations. Understandably, much of their politics, including the idea of creating separate states, has an ethnic slant. What makes conservation a practical objective in Bolivia is that most of the turmoil is taking place north of the parts of the altiplano that support the flamingo lagoons, which is, mostly, very sparsely inhabited. Where it is more heavily populated, as in Chile, a National Flamingo Reserve System is now in place.

Most important, there is now that dedicated international group of local scientists and resource managers, the Grupo para la Conservación de Flamencos Altoandinos, and they have a clear vision: an altiplano chain of interdependent laguna reserves, jewels in a unique belt of sanctuaries adorning the high plains of Argentina, Bolivia, Chile, and Peru with great flocks of exquisitely beautiful flamingos in breathtaking environments.

*That land is a community is a basic concept of ecology, but that land is to be loved and respected is an extension of ethics.*
Aldo Leopold

---

# 8
# *A Once and Future Patagonian Steppe*

"Deep Thought," the pensive computer in Douglas Adams's 1978 *Hitch-Hikers Guide to the Galaxy,* explained, "The answer to The Great Question . . . Of Life, the Universe and Everything is forty-two," an agreeably brief reply. However, several wildlife population biologists have concluded that the answer is closer to seven thousand.[62]

The "Great Question" that those scientists are trying to answer is, How big of a wild, free-living, adult population of a species is needed to ensure its long-term persistence? Consequently, how big should we make its reserves (as though we have much choice)? Their answer of seven thousand emphasized an approximate *interbreeding* population of adult vertebrates, not just the sum of out-of-contact population fragments. The number was reached by defining a minimum viable population (MVP) as one with a 99 percent probability of survival for forty generations and calculating the MVP for 102 species. The biologists considered population age structure, historical catastrophes, inbreeding depression, and demographic and environmental stochasticity (effects of random events on populations and habitats). Surprisingly, MVPs did not differ much among the species: The giant panda's was 6,224, the Asiatic elephant's was 4,737, and the bullfrog's was 5,909. Andres Novaro, working for WCS on reserve development and studying guanacos, wild Andean cats, and other predators, believes that the guanaco's MVP would be close to the 7,000 average.

Such an exercise is inherently crude—and thought provoking. For example, California's northern elephant seal's calculated MVP was 5,095, while that of the southern elephant seal (the species Bob, Bill, Kix, and I counted on Peninsula Valdés) was 31,709. How could that be? Well, different levels of risk. Almost all southern elephant seal populations are stable or declining, except for the one breeding on the Patagonian shore. Competition from fisheries is blamed. Rapidly increasing populations, even starting from a very small base, have good prospects—although the rate of increase had better not slow while the population is small. The northern elephant seal, now perhaps 180,000 strong, was once reduced, we suspect, to less than 20 individuals.

Of course, the *total* number of a species' population with 7,000 *adults* will be much larger because of all its nonbreeders, mostly juveniles. With some, such as flamingos, which don't breed until they are five or six years old, the total population might have to be four or five times the MVP, say 30,000 or more. (Besides, as pointed out in the previous section, flamingos are one of those highly vulnerable colonial species that *gather together to breed* in very few places. The loss of even one colony can halt the reproduction of a big proportion of the species.)

It remains difficult for planners to relate numbers to spaces, hence to carrying capacities, especially with regard to birds. Luis Jácome says that an Andean condor may forage over sixty thousand square kilometers, and even that is modest compared with the area used by an albatross. And what about the space-food relationship? Food is the burning question in animal society, famed ecologist Charles Elton observed, setting forth a crucial element for planning and managing wildlife reserves.[63] It is a prime factor in determining how sufficient a particular space really is from the animal's point of view and how large its numbers can be in that space, whatever its size—its carrying capacity. Moreover, every species has its own requirements. A huge space without cliffs is useless for barranqueros, one without lagoons is useless for flamingos and black-necked swans—habitat usefulness is species specific.

One might think that China's giant panda, with thirty-three reserves, would be secure. It isn't. The diminished numbers of the species are fragmented in approximately twenty-four populations. Currently, it is projected that the three subpopulations in the famous Wolong Reserve in Sichuan, each with thirty to forty-five animals, have more than a tenfold chance of extinction by 2100 if they remain isolated from each other—not interbreeding.[64] Most of the panda's isolated populations have fewer than fifty individuals. Though recent recounts provide some encouraging news about panda populations, the only way the panda is likely to long survive is through intensive management to minimize unfavorable events, or at least the effects of those events—whether inbreeding; disease; lack of food; poaching; or even bad weather, which might require artificial feeding. The same must be said for the whooping crane, Siberian tiger, California condor, Sumatran rhino, and the growing list of the walking wounded. The science is clear: Large numbers, protected areas, and responsive care are necessary for security. The less functional the ecosystem, the more intensive the necessary care. Unfortunately, science is only a small part of preserving nature.

In the days of the Tehuelches, there were immense numbers of guanacos on the steppes. There were Andean condors soaring overhead, crowds of choiques, mobs of maras, prowling pumas, flocks of elegant bandurria ibises (*Theristicus caudatus*), and, in the north, those patient plodding tortoises and raucous barranqueros. Where there was water, lakes were filled with flamingos, swans, grebes, coscorobas, and masses of brilliantly colored ducks. Working to recreate this marvelous pageant, at least in a few places, is an inspiring antidote to pessimistic acceptance of predictions that steppe wildlife will soon disappear. Almost every conservation scientist, every park ranger, and a great many schoolchildren now know what it takes to have wildlife: big enough, protected, cared-for habitat of the right kind. But it also takes a caring society and a functional, dedicated government (and conservationists prepared to help)—all that, plus ongoing scientific management.

Because Patagonia has so many recent immigrants, people are often undereducated about the wildlife and indifferent to its existence. There is little indigenous culture left, no gift of traditional ecological knowledge learned through experience by people who have lived there for centuries. Loyalty to the native wildlife and ecosystems is not common in the Southern Cone. Neither personal nor economic well-being seems to have been linked in most people's consciousness to the preservation of the more populous steppe environment, and certainly not to wildlife. European immigrants to Patagonia quickly imposed their traditional lifestyles and farming methodologies on the land, just as colonizers do almost everyplace. It was the familiar domesticated animals, especially sheep, that made life possible on the dry steppe, where most crops are impractical.

Perhaps that disconnect with native wildlife encouraged some of the irrational beliefs reflected in the ranching culture, that socially inculcated menu of half-truths and full beliefs that guides so much of its behavior. On one of my first trips to Patagonia, I was solemnly informed by a sheep rancher in Tierra del Fuego that one wild upland goose (which weighs about 6 pounds) eats as much as three sheep (whose combined weight is about 264 pounds). On my most recent trip, a provincial ranger in Neuquén explained that it was generally thought that one guanaco eats as much forage as three horses. (Actually, a 450-kilogram horse eats as much as four to six guanacos.) Such views are common in rural areas where native animals compete with livestock, and they make neither rational range management nor conservation easier.

Inescapably, the most common person-animal relationship on the steppes is rancher-livestock; all domestic animals are seen as resources to be used in some way, but also to be cared for. Wildlife is most often perceived as competitive, as a pest, a danger, or game. On the other hand, successful ranchers understand animal needs and the nature of predation and disease in ways most city people cannot. I speak feelingly of this connection, for I worked on ranches as a boy and am a zoo biologist who has cared for wild animals most of my life. The rancher's experience, if conservationists and wildlife managers gain access to it, could provide a foundation of useful knowledge for the creation of new kinds of sustainable reserves. It is also true, however, that not all Patagonian ranchers have paid much attention to native animals or to rangeland management, either, as the overgrazing of much of the steppe and the introduction of exotics makes clear. Some seem to dwell in a permanent state of ecological ignorance and denial.

Patagonians are entering a time of growing scientific knowledge and environmental stress, and also of public concern. But protection of steppes wildlife, and even good livestock care practices, is complicated by the fact that less than 5 percent of Patagonia's population actually lives on or manages the steppe ranches, although 95 percent of the steppes is privately owned. Enormous areas are subject to the will of a very small proportion of the population—a tyranny exists of small landowner and manager decisions often unrelated to each other or to the whole. Yet the sheep affects the guanaco, which affects the sheep, which affects the puma, which affects the fox, which affects the hare, which affects the choique, and so on. And climate change, new technologies, and economic shifts affect everything. Patagonian ranchers face a whole range of issues that cannot be adequately dealt with rancher by rancher without some larger strategy.

In the intermittently uncertain Argentine economy, there is little resistance to the blandishments of entrepreneurs pushing money-making ventures, no matter how ill considered. It is difficult for a shuddering populace, watching its leaders tiptoeing along a tightwire of national insolvency, to look ahead, to be concerned about the best long-term interests of their society and its environment. One need not be a sociologist to observe that people work first for family, only secondarily for society. In Patagonia, family ranching is a lifestyle issue even when it is not always economically viable. It is the same in much of the U.S. West. Only 20 percent of the income of small ranchers in the United States now comes from the land. Most have second jobs, and many have branched into tourism and services related to it. How do the sheepmen and goat herders on the relatively unproductive steppes perceive their future and that of their children? Can they be part of the solution for wildlife, or will they remain much of the problem? Can new wildlife reserves be designed that help them, that diversify their sources of income, and existing reserves be adequately cared for? Too often, Andres Novaro and Ricardo Baldi point out, reserves are likely to be condemned as breeding areas for livestock predators—a view that reveals little understanding of the origins of Patagonia's predator problems and one that will have to be explored in broadly participatory public meetings and studies. It will be extremely difficult to conserve wildlife if local people cannot meet their own needs. However, ranchers on the Argentine steppes enjoy three advantages that separate them from like people in many other regions: There are few people, there's a lot of space, and property rights are usually respected—except when they are public.

Hunting, like family ranching, is also a way of life in Patagonia, and a serious problem. Where guanacos and rheas are accessible, most are hunted out. The guanaco population has declined by more than 95 percent and the choique population probably just as far. The once amazingly approachable huemul, which formerly ranged over much of the steppe, has become one of the world's rarest deer and is found only in the southern Andean forests. Numbers of that model of monogamy, the mara, have drastically declined, and even the attractive mountain and plains vizcachas are now restricted to small pockets of rough country. Because hunting, along with the superabundance of predators supported by European hare, rabbits, and sheep and the depredations of feral pigs, has marginalized native animals, those who would bring back the wildlife of the original ecosystem are faced with complex tasks. They must reduce hunting of native herbivores, such as maras, guanacos, and rheas, while encouraging the scientifically managed hunting of native predators such as foxes, possibly pumas, and certainly introduced exotics, especially pigs, rabbits, and hares. If we reduce the hares and rabbits but not the predators, no choique chick or mara cub will stand a chance—not even guanacos will stand a chance in most areas.

The hunting of Argentina's exceptionally decorative ducks and geese is also astonishingly uncontrolled, and the Internet advertisements of its commercial purveyors (catering mostly to North Americans), which brag, "No legal limit," make startling reading. Here are excerpts from three:

> The birds will start flying before daylight, from their roost to the feeding fields. Average bag is 20 to 30 birds per hunter. After lunch and a little nap, the guests can choose to go trout fishing or shoot geese late in the afternoon.

> Since there is no legal limit, we suggest a self-imposed limit of 35 ducks per shoot. You'll enjoy both a morning shoot and an afternoon shoot each day.

> There is no limit in terms of quantity of geese to shoot per hunter. Normal hunt is 35–40 geese per hunter, going sometimes up to 100.

These people are shooting birds bred on properties hundreds of miles away, the public's geese. I have hand-reared the usual targets of these hunts from the egg, the beautiful upland (*Chloephaga picta*) and ashy-headed (*C. poliocephala*) geese, and find such ads indigestible. Nevertheless, sheepmen consider geese pests that compete with sheep for grazing, despite the fact that it takes fifteen geese to eat as much as one sheep.[65] Surprisingly, no one has yet claimed that geese have overgrazed the steppes! The nature of the advertised hunts, which have little to do with sportsmanship and much to do with the use of living creatures for target practice, reflects some of the less admirable aspects of nineteenth-century European and North American behavior, as well as Patagonian sentiment. But personal feelings aside, no population of large birds can endure uncontrolled slaughter. Not even North America's extraordinarily abundant passenger pigeon, which may have numbered as many as nine billion, could withstand it. Millions were slaughtered for sheer sport, like Argentina's geese, and their carcasses fed to hogs. In 1914, the last one died in the Cincinnati Zoo. Up until the mid-1950s, there was a particularly beautiful little goose in southeastern Patagonia, the ruddy-headed goose (*C. rubidiceps*). It was hunted, its nests destroyed, and then foxes, introduced to kill introduced rabbits on Tierra del Fuego, did the rest. It is now virtually extinct in Argentina, its only viable population in the Falkland/Malvinas. For the most part, Argentines do not base or enforce hunting kills on species population assessments. Unless people insist that they do, they won't.

It is not that Patagonians lack a strong conservation ethic, but that the majority of those who have higher education levels and care about wildlife protection are townspeople, disconnected from the control of wildlife—which is almost a worldwide generalization. It seems inevitable that some of the desire to protect nature that does exist will have to be enforced by government. Had the preservation of Yellowstone National Park been left to a few Wyoming ranchers and developers, I wonder if there would be a park.

Fortunately or unfortunately, almost all of the steppelands are in Argentina, 750,000 square kilometers (about twice the size of Germany). There are no multinational complications. So Argentines, and only Argentines, will preserve them or destroy them. Development is patchy, and it is impossible to say where it will focus next. We know, however, that development almost always brings environmental degradation and that its benefits, which may be substantial, are very unevenly distributed. Oil companies in Mendoza and Neuquén make large contributions to the provincial governments and, consequently, enjoy nearly total control over whatever lands they wish, completely occupying Auca Mahuida, one of Neuquén's most important reserves. One wonders how much of their income will stay in Patagonia and what the provincial governments will do when the oil fields are exhausted.

Some insist that conservation "developments," by which is usually meant such minimally invasive entities as parks and the recreational facilities they attract, must also provide major income, which would not be much different from asking police and schools to provide major income. Security, education, and conservation are always publicly profitable but only sometimes privately so. They just happen to be essential to civilization. Nevertheless, conservationists in Patagonia, as elsewhere, often seek to justify preserving environment and wildlife by citing the tourist revenues such preservation could bring. In the

U.S. West, it is a fact that counties with protected public lands are becoming the most economically diverse and prosperous.[66] It may well be that steppelands tourism can produce new, long-term income, if managed with imagination, but not in comparison with the temporary oil business. Ultimately, preserving nature in parts of the steppes must be for Argentines and for forever, not just for tourists and for now.

The unhappy outlook of Mr. Darwin's ostrich population, so deeply diminished and threatened, the choique barometer, confirms that nature on the steppes is sick. If its diagnosis is doubted, we have additional prognostications from the guanaco, the mara, the tortoise, the barranqueros—and from the scientists studying them. Protection that supports only small numbers of guanacos, maras, vizcachas, black-necked swans, and choiques here and there in fragmented populations will not secure their survival, as we've seen. A sobering example comes from the research of Andres Novaro: Guanaco populations of less than eight per square kilometer, he finds, are simply not viable in the face of significant puma predation, while those with more than eight per square kilometer have a good chance of surviving. It will take a policy of active management, as well as passive protection, to secure Patagonia's wildlife, and only large, scientifically supervised reserves may be able to do the job. Conservation policies are not primarily set by scientists, however. They are set by politicians, by departments of economic planning, industry, agriculture, energy, and trade, and by that diffuse tyranny of thousands of individual decisions by people not necessarily considering any of the concerns of the other entities.

Auspiciously, wildlife experts in every Patagonian province are developing long-term plans and priorities, but there needs also to be a shared vision for the whole of the steppes, a "Grand Plan." Andres Novaro and Ricardo Baldi are working closely with provincial experts and promoting such ideas in an attempt to consolidate certain protected areas in what could become a "Patagonia Reserve Network." Ultimately, such planning must involve many "stakeholders" and seek the best ideas.

At the beginning of this chapter, I introduced Deep Thought to provide a precautionary note of complexity, to suggest some of the background science necessary to create lasting nature preservation, whether in the guanaco's desert steppes or the panda's bamboo forests. Fortunately, both biologists and far-sighted officials are giving the embodiment of their "dreams" much thought, so I have also encapsulated some of those reveries below. The strategies I've heard and dreamed about for steppe and altiplano regions include three main elements: I'll call them Tehuelche Landscapes, Species Sanctuaries, and Wetland Refuges.

First, imagine primary reserves big enough to sustain core populations of the steppe landscape species—guanacos, rheas, and maras—as the Tehuelches knew them: "Tehuelche Landscapes." Without help, they will attract the natural predators: pumas, wild cats, and foxes. Manage each reserve to sustain ecologically compatible populations, including unpleasant control and culling. Rigorously remove aliens—plants as well as animals—and focus on appropriate numbers of each native species in relation to carrying capacity. Monitor health on a ongoing basis, a necessity made clear by recent eruptions of disease (e.g., Ebola, SARS, rabies, West Nile virus, deer wasting disease, and hantavirus in Patagonian rodents) in far-flung locations. Finding land for such great landscape reserves is challenging, although Payunia, mentioned earlier in relation to grazing cattle, is a superb possibility. Were it connected with Auca Mahuida and Llancanelo, a great, biologically viable, megareserve could be created. (Andres Novaro, Susan Walker, and Ricardo Baldi calculate that it would cover eight thousand square miles, making it 2.33 times as large as Yellowstone.) In some places, local support for truly sizable Tehuelche Landscapes might be won by including them within a ring of "extractive reserves," managed not only for conservation and tourism but also to provide economically attractive numbers of guanacos for shearing or harvesting through contracts with local ranchers.

Species-focused preserves are the second element. Visualize "islands" of species-specific habitat—barranquero cliffs, bandurria ibis colonies, mara and Chaco tortoise concentrations—perhaps separate from Tehuelche Landscapes but managed so as to ensure the survival of especially vulnerable species that require specialized habitats: "Species Sanctuaries." A barranquero reserve at El Cóndor, a mara reserve on Peninsula Valdés, a tortoise sanctuary near San Antonio Oeste provide examples.

Finally, there are the critically important wetlands, the most biodiverse of steppe habitats, places like Lagunas Tromen and Llancanelo. If not included in Tehuelche Landscapes, they would be placed so as to sustain spectacular populations of aquatic birds: "Wetland Refuges." These flamingo, shorebird, and waterfowl habitats are rare in Patagonia and easily degraded. Yet they sustain the greatest numbers of species and many of the most vulnerable.

I imagine that most steppe reserves, given the cherished independence of the provinces in Argentina's political system, will be provincially administered, but they can fulfill the needs of wildlife preservation only if they are part of a steppes-wide interactive program. Reserve management must be cooperative, based on science, and subject to regular public review. For such concepts to work, two things must happen: First, some private landowners must buy in, presumably from an extractive reserve, harvest, and tourism point of view, for sheepmen own the vast majority of the steppes. Second, reserves must be managed to fulfill their biological functions—not simply protected from poaching but developed to fulfill the feeding and breeding needs of their native wildlife. This is not just police work. The challenges range from creating waterfowl breeding areas in degraded wetlands to balancing predators, prey, and native plants in larger, often overgrazed steppeland habitats.

Compared to wetland protection in the steppes, laguna protection in the Andean altiplano seems daunting. Besides the fact that four nations are involved, the poverty of many of the people and the activity of large mining companies, with no attachment to the land or its wildlife, in much of the region complicate matters. However, few people live at such high altitudes, and, most encouraging, Grupo para la Conservación de Flamencos Altoandinos is hard at work promoting an extraordinary multinational chain of interdependent laguna reserves, a concept already receiving both government and private interest.

The as yet unvoiced ideas for the steppes are no less powerful: the restoration of its once unique world-class wildlife spectacles—which means what? Say, 2,000,000 guanacos! If there are now 500,000 guanacos in Argentina, the goal could be 2,000,000 or more. With such numbers, ranchers participating with Tehuelche Landscapes could harvest or shear far more than anyone does today and without threatening the populations. The harvesting (awful term—they are not potatoes!) could profit ranchers whose land borders the reserves and helps support the wildlife, not weekend or commercial hunters from the towns, unless invited by participating ranchers. Darwin's rheas are a special case. They have few detractors, and their feeding habits are not competitive with sheep; but there are no believable population numbers for these strange creatures—there's no starting place. In spite of that, let's set a target pending investigation. How about 4,000,000? That would be a serious population, one that could be "harvested," if that is really necessary to win local support, and one that could help make Tehuelche Landscapes an exciting national treasure and global phenomenon.

---

The majority of the Southern Cone's most spectacular wildlife does not live on the steppes or the altiplano, but along its still largely pristine Atlantic coast. There, the wildlife population is benefiting from fledgling conservation efforts, international tourism, and a remarkably collaborative Patagonian Coastal Zone Management Plan now being developed. But coastal wildlife has a highly vulnerable and more complicated relationship with people than the wildlife of the steppes does. Exactly what that means is a very different story.

Coastal Patagonia and the Falkland/Malvina Islands

# *Coastal Chronicles*

*The two thousand miles of beaches and cliffs that run from the Rios Negro and Colorado to Tierra del Fuego are home to an assemblage of big, appealing, intensely colonial creatures, whose life stories define the interface between land and sea. Because of their numbers and sheer weight (their biomass), the amounts of other creatures they eat, and the distances they travel, they are ecological superpowers, affecting both the ecology and the human economy of the long Patagonian coast of the southwest Atlantic.*

*The big seals and sea lions of the shores were also slaughtered—like the guanacos and rheas of the steppes—but almost to disappearance. Unlike the wildlife of the steppes and the remote altiplano, they now enjoy increasing public interest, yet they are threatened anew by overfishing of their foods and by shoreline developments.*

*The shore is an ancient world, for as long as there has been an earth and a sea there has been this place of meeting of land and water. Yet it is a world that keeps alive the sense of continuing creation and of the relentless drive of life.*

Rachel Carson

---

# I
# *The Doctor and the Sea Lions*

It was late in the cool austral spring along the same Peninsula Valdés shore where I had counted elephant seals eighteen years before, fifty kilometers north of the WCS station where Graham lived. Huge, muscular, bitter-chocolate–colored sea lion bulls had staked out territories at Punta Norte's high tide line and were competing to gain and retain females for their harems. Graham and I left our rented Ford Falcon (manufactured in Argentina long after its extinction in the United States) on the road and scanned the sea's edge with binoculars. The wind was brisk, the surf high, and the cloud cover solid. Tehuelche pebbles covered the shore and were warm to the touch, hoarding radiation from a stingy sun. They were interrupted here and there by the dry gray bones of sea lions slaughtered years before. Giant petrels (*Macronectes giganteus*)and black-chested buzzard eagles (*Geranoaetus melanoleucus*) wheeled in the wind overhead, while red-beaked oystercatchers paced the water's edge in pairs so close together they seemed invisibly attached. The pristine beach extended south as far as the eye can see, toward Caleta Valdés, Punta Hercules, and the cliffs of Punta

Delgada, thence around Peninsula Valdés' southern corner to Puerto Pirámides on Golfo Nuevo and the town of Puerto Madryn at its base.

To the east, we spotted a particularly formidable-looking sea lion bull staring down at a slender human figure lying flat on the beach close by. The bull's authoritative gaze was backed by seven hundred pounds of muscle and huge canines. Finally sensing our presence, the human about-to-be-a-statistic inched backward, getting far enough from the sea lion that he could retreat in a chastened scuttle and come our way. He was tall and lean, with a short black beard and clear hazel eyes, and wore a fur-hooded, down-lined jacket, tennis shoes (no socks), and shorts—dress that seemed peculiarly appropriate to a place where temperatures swing from 50°F (10°C) to 100°F (38°C) at this time of year. He had a battered notebook, bruised binoculars, and a marvelously conspiratorial grin.

"Mi nombre es Claudio Campagna."

After six years of medical school, Claudio had become a physician living in Buenos Aires, with a promising career—and a serious problem: He did not want to be a doctor. He had discovered a different calling, an obsession; he had fallen in love with sea lions, among other marine mammals. He is irrepressibly enthusiastic about sea mammals, about Patagonia, and about Peninsula Valdés. He has a consuming passion for science and wildlife. He explained to us that he wanted to pursue a doctorate in animal behavior and work on the conservation of the Patagonian coast—and lost no time in describing his dream.

It was 1982 when I met Claudio, and his plan to study with animal behavior experts in England went up in smoke when Argentina attacked the Falkland/Malvinas. Nevertheless, five years later, he had added a Ph.D. to his M.D. at the University of California (Santa Cruz). Working under the supervision of the distinguished marine mammal expert Burney Le Boeuf, he completed a dissertation on the breeding behavior of the southern sea lion.[67] WCS supported part of his research and travel costs, as did Argentina's national science foundation, CONICET, but Claudio paid most of his own way by teaching anatomy at the medical school—a challenging birth for a future to be plagued with economic worry. He was moving from the certainty of being a needed physician to the uncertainty of being a wild animal conservationist in a place that wasn't sure it wanted one. Since then, he has not only managed to keep faith with his beloved sea lions and Peninsula Valdés but also become an advocate so moving that he brings lecture audiences to tears.

Sound information knowledgeably yet passionately interpreted is the front line in the battle to win hearts and minds to save wildlife. It's a weapon that's hard to disable. If the information is fascinating, so much the better. Claudio's doctoral research asking questions of sea lions is an example. Southern sea lions (*Otaria flavescens*) are large, formidable relatives of the California sea lions that live along the U.S. West Coast. They breed from December to February, and there are about fifty colonies, loberías, along the Patagonian coast. Patagonia's elephant seals are far larger, but their elefanterías are confined to Peninsula Valdés, and their diet and breeding season are almost entirely different. Both of these big predators are major actors in the Southern Cone's ecological drama. Their stories tell us a great deal about many of the other animals with which they share the coast. If they can truly thrive there again, chances are that much else can, too.

Sea lions are called *leones*, lions, or, more often, *lobos*, wolves, in Argentina, and about 460 of them were breeding at Punta Norte when I met Claudio. He had begun his studies by working out the main details of their reproductive behavior. To learn about it, he spent hour after hour hiding in a blind among bushes at the back of the beach and recording every nuance of behavior, comings and goings, interrelationships, fights and near fights, births, and disappearances.

Females give birth three days after arrival at the breeding areas, copulate once six days later, for about eleven minutes, and go to sea for the first time again two days after that. Thereafter, the mothers alternate periods of feeding themselves at sea (about 2.8 days at a time) with periods of nursing their pups on land (about 2.3 days at a time). Males corral whatever females they can and fiercely defend them from other males. Each bull seeks exclusive access to as many receptive females as possible, which is not many in this powerfully competitive society. Young bulls up to about eight years old, weak bulls, and old bulls past their prime cannot win the central harem sites most favored by females, or hold on to harems. Those losers must live on the fringes of the rookery.

As the breeding season approaches, the big males are the first to appear on Punta Norte's empty beaches. They arrive one at a time and stake out territories where they expect the females to show up, in what Claudio calls the Central Breeding Area. The territories they claim are only about forty-five feet in diameter. Those I saw later in the season, squeezed by competition, were much smaller and were lined up like psychological fortresses along the high tide line. Harem bulls seem to come back to the same areas year after year; Claudio was able to recognize, from scars, two males that defended precisely the same territories on this featureless beach two years in succession. Although the females arrive later, they know where the Central Breeding Area is and either agree on its suitability or simply head toward a place where they see the densest concentrations of males.

As if competition were not enough, sea lions at Punta Norte must also deal with problems of temperature and tide. Air temperature on the beach often exceeds 86°F for four to six hours a day during January, the prime breeding period. This is truly uncomfortable for big animals adapted to very cold water, where thermal conductivity is twenty-five times that of air. At times, air temperature rises to 103°F. Twice a day, males, females, and pups get some relief as high tide comes up to them, yet none deliberately enter the water to cool off.

The prime bulls do not space themselves out at polite distances, as some pinnipeds do. They live scarcely a lunge apart, and they have to get along with only two to four females in their harems, not the twenty or more common in California sea lion harems. To retain females, the bulls must be there all the time, and they are, without feeding, rarely moving more than a few feet from their harems, day and night. I found myself cringing when a huge male, easily three times the weight of his mates, grasped one of the females in his teeth and literally threw her back home when he considered she was straying from his territory. Several females had bloody wounds on their necks as a result of such behavior. Unlike the coarse-looking males, the females are graceful and a wonderful golden color, altogether beautiful animals. The payoff for the males was clear: Ninety-four percent of the males that held a territory for five or more days copulated, and the first males to arrive at the rookery had mean tenures of twenty-four days.

Standing with heads high and arched back over their massive necks, aggressive males make explosive staccato bellows and wet "chuh-chuh-chuh-chuhs"—ad infinitum. They have nothing like the deep, booming calls of bull elephant seals. To me, they sound like a four-cylinder truck running slowly on three cylinders. Real injuries can occur in the male sea lion rushes, retreats, and fights with neighbors. One of my photos shows a thumb-sized canine tooth in midair after being broken out of a fighting bull's mouth. Gaping cuts in necks and chests and sometimes serious eye injuries occur. By the third of January, Claudio recorded that bulls were having as many as ten "agonistic interactions" per hour. As the season progresses and more males win positions in the Central Breeding Area, the herding activities of males trying to retain receptive females seem more desperate—they have been a long time without food.

In the meantime, the harems fill with impossibly fragile-looking newborn pups. Actually, they weigh nearly twenty-five pounds, but the adult males and even the females are so large that the newborns appear tiny. As I watched a birth, four giant petrels landed and began tearing at and gulping down the bright pink placenta. They were followed by a half dozen kelp gulls, which discovered the pitifully flattened body of another pup, probably crushed in the frequent scuffling and charging of the bulls. It is a gory scene. The ravenous petrels fight over the bloody placentas with huge, hooked beaks under the very noses of the sea lions. Kelp gulls dive headfirst into the melee and emerge with their snow-white breasts streaked with blood. Female sea lions snap at the birds as they protect their newborns, while snowy white sheathbills walk fastidiously on the outskirts of the uproar snatching overlooked fleshy tidbits.

Basic to wildlife study is finding some way to keep track of individual animals. At the beginning, Claudio could not distinguish more than a few of the lobo bulls and females, apart from those with obvious scars. He saw sea lion faces in his sleep but no easy solutions. Animal identification and tracking techniques range from simple leg bands for birds to internal physiology monitors stuffed down a snake's mouth, and to sophisticated radio satellite tags. Biologist and WCS director for science George Schaller found that he could tell mountain gorillas apart by their noses and made a nasal gallery of all the animals he could see. Iain Douglas-Hamilton, working with African bush elephants, made full-front diagrams and photos, which concentrated on ear rips and tusk differences. Tony Lynam in Asia and Alan Rabinowitz and Carolyn Miller in Central America used camera traps to photograph tigers and jaguars from relatively close by because the stripes and spots of the cats are as distinctive as fingerprints. Sea lion variation is more subtle. The selective demands of swimming and catching food in a cold, watery environment, the very physics of that environment, have acted to homogenize and streamline, ironing out easily visible individuality — no spiked hair or belly-button studs here.

Finding a nondisruptive way to mark the sea lions was Claudio's fundamental problem. He finally decided to use splattery paint pellets, whose marks would be distinguishable for three to six weeks. They would be the least unsettling of the few options offered; he would not have to capture the animals, but neither could he run up behind them and paint their bottoms, as Graham did with peludos. He settled on soft, marble-sized paint pellets that could be projected at the sea lions with a slingshot or $CO_2$ pistol. While his accuracy with the slingshot bore testimony to a misspent youth, the air pistol proved better because of the stalking problem. Using a slingshot, you must handle the soft paint pellet very carefully so as not to break it all over yourself rather than the sea lion. A $CO_2$ pistol could be loaded more cleanly and aimed while the scientist was flat on his belly.

Unfortunately, a "gentle" $CO_2$ pistol has only a short range; Claudio didn't want to use a more powerful gun because it might bruise the animals. There was no alternative to wriggling very close to his quarry, preferably within thirty feet. This was tricky work. "Painting" had to be done without the animal seeing where the blob originated. Nevertheless, thirty females and sixty males were successfully marked each year for three years, and I crawled behind Claudio, photographing the process, on several occasions. The sea lions sometimes flinched when the pellet splattered on their chests or shoulders, but that was about all. The result was a gallery of colorfully spotted sea lions and "personnel" cards faithfully recording each splatter pattern.

But the Jackson Pollock approach to marking sea lions is not without hazards. Using an air gun to propel a blob of paint in a strong wind is a prescription for self-decoration, and Claudio and his colleagues were sometimes more distinctively recognizable than the animals. Because darkness provided better opportunities for getting close, he often stalked sea lions at night, when the wind was still and he could see his targets silhouetted against the sea. Only morning could reveal precisely how they were marked, however.

Despite the scientist's recognition problems, sea lions (like geese, elephants, zebras, flamingos, crocodiles, hunting wasps, and much else) have no trouble recognizing each other. I don't mean that all the members in a herd can identify each other individually, but members of family groups and neighbors in colonies certainly know each other. We commonly underestimate that ability and the ability of animals to recognize us individually. Keith Kendrick and his colleagues at the Babraham Institute near Cambridge worked with a group of twenty sheep and taught them to recognize frontal views (photos) of the faces of twenty-five pairs of their fellow sheep by associating one of each pair with a food reward. The sheep were considered to have "learned" when they chose the reward face 80 percent of the time. They could still recognize the pairs up to eight hundred days after the training period. Moreover, they could distinguish among profile views of the same faces without being retrained; they could also remember human faces.[68]

It is important not to underestimate animal abilities as you try to anticipate what the creatures you are studying and trying to conserve will do. They are more sensitive to what is going on around them, and considering much more of what it means to them, than is often thought. Their interrelationships are a part of what they themselves *are*, and it is useful to start with the assumption that they know each other, trust or fear their neighbors, or dominate them, and respect or ignore the "individual spaces" of others. These are definitive, if non-visible, attributes of each animal.

Claudio's southern sea lions have enormous front limbs that have been modified as flippers but are still capable of considerable mobility. I have watched the zoo's California sea lions scratch their sides and the back of their heads with their hind flippers, as a dog would with its hind foot, and I have seen trained animals cover their faces with their front flippers, catch balls, and, silly business, wave with them. (While trained animals may do lots of things that they would not do in nature, they won't do anything that they cannot do in nature.) A sea lion's back flippers, unlike those of true seals such as the elephant seal and the harbor seal, are reversible. The creatures can bring them forward like the legs of other four-legged mammals and gallop on land as rapidly as a man can run (although not as long), which is something worth remembering. It is rumored that two people have been killed by southern sea lions when they got between the sea lions and the water. Sea lions have tiny, flaplike ears and big, protuberant eyes that constantly produce masses of rinsing tears, to make up for their lack of the usual kinds of eyelids and eyelashes. Their eyes are capable of gathering a great deal of light underwater as well as distinguishing acquaintances on land.

Sea lions and fur seals (*Arctocephalus* sp.) differ in the view of many people only in the way set forth by their Argentine names. The sea lion is called *lobo marino de un pelo* (sea wolf of one skin) and the South American fur seal *lobo marino de dos pelos* (sea wolf of two skins). Both were devastated by sealers. The hunt for the latter's relative, the Antarctic fur seal, was the largest for any pinniped—millions were killed. In Argentina, however, the South American fur seal's preference for remote rocky islets made them comparatively scarce and inaccessible.

A sea lion, with *un pelo,* has a fair covering of fur but sparser than that of most land mammals. A thick layer of fat under the skin provides insulation and streamlining and supplements the thin fur. Its skin is primarily used by sealers for leather products. The fur seal has comparatively thick fur. It consists of two layers, *dos pelos,* protective outer guard hairs and extremely fine and thick underfur, which remains dry by trapping tiny air bubbles in its oily fibers; the seal also has a thick layer of fat under its skin. Its pelt is used for fur garments and is much more valuable than that of the sea lion. Fur seal underfur contains about three hundred thousand hairs per square inch, while the human head (of the fortunate among us) has an average of only about one hundred thousand. (Both are put to shame by the sea otter's fur, which can contain up to one million hairs per square inch.)[69]

The largest population of South American fur seals in the southwest Atlantic today, about 250,000, is in Uruguay, where it has been harvested at least since 1515, mostly under controlled conditions. A big male is about half the size of a big sea lion bull. With the decline in the use of seal furs, the Uruguayan hunt annually takes about 5,000 males, primarily for their testicles, for the Asian aphrodisiac trade.[70]

Imitating a molting elephant seal by looking scruffy and crawling on my belly, I have slithered over Patagonia's beaches countless times to photograph sea lions. The secret is to keep flat, as Claudio was when I first saw him, and move slowly, but such a stalk is potentially dangerous; and, if done carelessly, it is disturbing to the harems. Besides, when the bulls fight, they disregard everything about them and charge heedlessly over nearby objects. After a couple of close calls, I bought a more powerful telephoto lens.

Southern sea lions differ in an exceptionally violent way from three of the four other kinds of sea lions in one of the methods they use to sequester females. Groups of young bulls, unsuccessful at establishing harems in the Central Breeding Area, may try to steal mates—acting as a group. They congregate on the outskirts of a colony, a frightening-looking pack of young toughs. Then they swim quietly en masse through the surf below the Central Breeding Area and, suddenly, make a gang charge up into it, attempting to overrun the "beachmasters" and seize females from their harems. Such a raid is a wild battle, unlike anything else I have ever seen.

Imagine the scene I once witnessed: Huge harem bulls, largest and strongest of their tribe, are at the water's edge, each bull with two to four females accompanied by tiny black pups squalling for attention. Without warning, erupting from the surf like giant frogmen on a demolition mission, a surging swarm of glistening, roaring, landless males appears. Typically, there are ten in the raiding party, Claudio has found, but on this occasion, there are fourteen. Eyes rolling balefully, mouths agape with rasping growls, they rush toward the center of the rookery trying to isolate and corral females—a surreal wetsuit commando. Pandemonium breaks out. Chaos!

Response from the harem bulls is excruciatingly slow. Not one is willing to act unless his own harem is threatened. But the raiders' advance overruns the turf of three bulls at the same time, and the threatened beachmasters seem to swell in anger, wheeling to defend their harems. They counterattack with a ferocity they lack in ritualized skirmishes with their neighbors. Their roars overpower the sound of the surf. The breeding bulls literally crash into the raiders, pebbles spraying into the air, head-on collisions sending shock waves down thickly muscled seven-hundred-pound bodies. In effect, if not by intent, the harem bulls fight in concert, side by side, and with all their strength, charging, blocking, biting, and bellowing. Although their counterattacks can seem somewhat cooperative, each bull defends his own females, and, once the raiders penetrate the breeding area, they fight as individuals. The raiders compete among themselves in a chaotic situation of everybody against everybody. The females try to protect their newborn pups, which are ignored, trampled if in the way—usually emerging unscathed. A Patagonian sea lion raid is reproductive war, a battle to procreate, and for drama, it has few wild animal parallels.

As I crouched in Claudio's blind, one big raider overran the defenders and, once behind them, grasped a large female in his teeth, hoisting her high above the fray—she grimaced as he held her in his cavernous mouth between inch-thick canines—and then throwing her from the beach into the surf, where he rushed after her. Another raider pursued a female up the beach, causing the harem bull, in hot pursuit, to leave his harem undefended and allowing a younger bull to run off with a newborn pup, a strategy that sometimes brings its mother, too.

As suddenly as it began, the raid was over. One pup was seriously injured in the battle, but no raider gained a site or won a female in the

Central Breeding Area. I had taken more than a hundred pictures—and, unscientifically, rooted for the defenders.

During the peak season, January 14–26, raids occurred once during every two hours of observation. Claudio calculated that two-thirds of the raiders were subadult males (less than eight years old) and that in about 50 of a series of 355 raids, a raider successfully remained in the Central Breeding Area with one or more females. So raiding is a more successful mating strategy than it appears.

Are female sea lions really only passive pawns in the males' breeding strategy? "Not entirely," said Claudio. Paint marks had enabled him to discover that the females, the *hembras*, may move from harem to harem, though usually only during a raid or after estrous, when they are of no reproductive value to the beachmaster—"Maria" had visited three harems in one day.

But an understanding of sea lion lives, loves, and child rearing on the beach tells us little about what they do at sea. What do they eat? Where do they go to eat? What eats them? What is their relationship with the growing commercial fishery along the Patagonian coast? It is in their search for food at sea that their future is most uncertain.

Time-depth recorders (TDRs) and radio satellite tags are the investigative tools of choice for tracking animal behavior at sea, but to attach them to sea lions, the animals must be anesthetized, with all the worries of possible injury that that implies (to humans as well as sea lions). WCS has a field veterinary staff with unparalleled experience in handling formidable wild creatures safely. Elephants, rhinos, gorillas, lions, cheetahs, crocodiles—a vast diversity of wild creatures have been patients either in the zoo or in the wild, but sea lions pose special problems.[71]

Darting a wild sea lion with an anesthetic is not hard, but there is a danger that the animal will run into the water before the anesthetic takes hold and then drown when it does. Besides, the choice and amount of anesthetic are complex decisions, and reactions to it vary. Females were Claudio's first targets, but that meant that the darting had to take place during the few days between pupping and the female's return to the sea to forage. The idea was to catch a harem female and her pup, anesthetize her, do a quick health examination, getting blood and stool samples, attach a tag, bring her out of the anesthesia, and put her and her pup back into the harem without injury to pinniped or scientist. This was another kind of raid.

Sliding a long pole with a lasso-like loop on its tip before him, Claudio wiggled closer and closer to a harem high on the beach. He gingerly placed the noose over the nearest baby and pulled it gently away. The alarmed mother followed, was quickly noosed herself, wrapped in a net, pulled up the beach, and anesthetized by a small team slipping and sliding on the shingle. They moved so fast that it all happened before the harem male could make up his mind to charge.

Once the female was anesthetized (veterinarians Billy Karesh, Bob Cook, and Marcela Uhart eventually developed a portable gas anesthesia system for the purpose), she was constantly monitored during the health evaluation and attachment of the recorder. One of Claudio's students poured cool water on her flippers to be certain that her temperature did not become dangerously elevated in the desert sun, while another mixed epoxy adhesive. Claudio affixed the TDR or satellite tag with the adhesive to the fur on the back of the female's head, with the fervent hope that she would not lose it before it could be recovered. TDRs are relatively cheap, but at $3,000 to $4,000 each, satellite tags cannot be deployed casually.

With everything in place, the sea lion baby (which usually accepts a member of the human team as "sitter" during the handling of its mother—in the instance I witnessed, tall, un-sea-lion-like Rodolfo Werner) was placed back with its mother, and the medical staff reversed her anesthesia. Recovering, Mom looked around with the air of someone who has just been mugged but found nothing missing and made her way back to the harem with her pup. I was surprised that tagged sea lions did not attempt to rub the little contraptions off or even seem to notice them.

In an examination of 18,057 sea lion dives, Claudio and Rodolfo Werner found that most lasted only two to three minutes, at least those of lactating females. The majority were between 19 and 62 meters (62–203 feet) deep; the maximum depth reached was 175 meters (574 feet). These sea lions, especially the females, were fishing on the relatively shallow continental shelf. They dove at the rate of eleven to nineteen dives per hour, with no real difference found between diving rates by day or night. Biologists in the Falkland/Malvinas, however, have found that most diving by sea lions is done at night (and one tagged sea lion dove to a depth of 243 meters, or 797 feet), when they must be able to see little but bioluminescent creatures and be using other senses.[72] Totally blind sea lions have been found apparently healthy and well fed. Intercepting fish underwater when you are blind seems comparable to catching a base hit by Joe DiMaggio with your back turned—but there is more to it.

Dolphins and whales have evolved echolocation capabilities, which allow them to sense the outlines of their surroundings and even those of their prey. Bats locate flying insects at night with analogous sensibilities. Now a group of German scientists has discovered that at least part of the fish-catching ability of seals (and sea lions, we expect) depends on their whiskers. They employ their long vibrissae not only to sense the size and shape of objects close up but also to recognize tiny movements of water, such as the wake created by passing fish or invertebrates. Blindfolded harbor seals could "use their whiskers to detect and accurately follow hydrodynamic trails generated by a miniature submarine."[73] Moreover, most pinnipeds have excellent hearing underwater, and their ears are modified to improve directional hearing.[74] They also have good memories. A University of California study found that a female California sea lion that in 1991 had learned letters and numbers—"symbols"—and also the concept of "sameness" in symbols was just as good at recognition tests in 2001, though she had had no chance to refresh her memory in the intervening years. So it is reasonable to suspect that sea lions can remember not only their friends but also where they have been fishing.

Satellite tagging enabled Claudio and his colleagues to develop maps showing precisely where the sea lions had been—basic information for conservation. They were revealing the seascape according to sea lions. An exceptional female, "N5," foraged 161 miles (259 kilometers) from the rookery, and one of her trips lasted nine and a half days, during which she covered at least 540 miles (869 kilometers) without any "haul-out" on land. However, as baseball's Casey Stengel once said, "They say you can't do it, but sometimes it doesn't always work." Once, one of Claudio's satellite tags reported its sea lion swimming 150 miles west in the Patagonian desert.

Highly efficient foraging is critical to the ability of breeding males to fast for two months or more while defending their harems. The few records Claudio and his colleagues have collected so far suggest that the males swim farther and dive deeper than the females; their world is bigger.[75]

Dissection of Patagonian sea lion scat proves that the females eat at least thirteen species of fish, five species of squid and octopods, and three species of crustaceans. The main prey of the females seems to be the red octopus (*Enteroctopus magalocyathus*) and the Argentine shortfin squid (*Illex argentinus*), while males forage mainly on Argentine hake (*Merluccius hubbsi*) and Patagonian squid (*Loligo gahi*).[76] Their total catch includes many fishes and other creatures of little interest to commercial fisheries, but their food choices do overlap with those of the fisheries, especially with regard to hake and squid, as do the catches of many other species. For example, dusky dolphins favor exactly the same sizes of hake used by the Argentine fishery, while Commerson's dolphins take hake of smaller sizes. But, despite all the work of Claudio and other pinniped biologists on the Patagonian coast, there is still much that we do not know. An obvious question is, Why are there 68,000 sea lions rather than 680,000?

It is relatively easy to gather up a gang of "bully boys," tramp down the beach, and club sea lions to death for their skins (or penguins or albatrosses for skins and feathers). It takes no thinking to kill them until there aren't any more and then do something else. Suppose, however, that you want to preserve sea lions. It takes layer after layer of training and planning. It cannot be done in a day, a week, or a year. It takes countless hours of observation, comparison, analysis, presentations, and arguments with those who simply wish to kill them or use their rookery area for something else. It takes ongoing responses to critics, along with teaching and even begging—and finding money to do all those things and to keep doing them. It takes the development of a local constituency for sea lions, and its maintenance.

Even then, there is the worrisome question of the sea lion "disappearances." Sometime between the late 1930s and early 1990s, an enormous southern sea lion disappearance took place in the Falkland/Malvinas Islands—300,000 or more vanished. To this day, no bodies have been found, and plausible theories about what happened are in short supply. The biggest of the sea lions is the Steller's sea lion of the Bering Sea, and it is now disappearing. Overfishing is suspected. The Steller's sea lion has been given "Emergency Threatened" status by the U.S. Department of the Interior.

In the late 1930s, there were about 380,000 lobos in the Falkland/Malvinas, but by 1990, scarcely 3,000 were left. While it is true that some 40,000 were killed for their skins between 1928 and 1938, few were taken thereafter. Major commercial fishing efforts in the area had not yet begun, so competition with fisheries does not seem to be the explanation. Something else affected either the sea lions themselves or the area's carrying capacity for them.

Orcas (killer whales) are as much a fact of life for Peninsula Valdés sea lions as African lions are for wildebeest and wolves for elk. In recent years, orcas have been taking five hundred or more sea lion pups each breeding season. They are devastatingly effective predators and could be looked upon as a natural "disease" of sea lions. Orcas have been observed herding sea lions into tightly packed groups while individual whales take turns charging among them and eating the trapped animals. Penguin expert Jory Wilson saw a pod kill over twenty in an hour. But Patagonian orca numbers are small, and their presence is not new—why would an age-old predation suddenly depress sea lion numbers? Some biologists nevertheless think orcas are the explanation of the Falkland/Malvinas' sea lion disappearance. They suggest that the sea lions disappeared in the wake of a period of intense whaling, which reduced the orcas' main prey and caused them to turn to sea lions.

What about disease? In 1988, the seal population of northern Europe, mostly harbor seals (*Phoca vitulina*), was devastated by a distemper epidemic. About eighteen thousand died from that morbillivirus. In May 2002, sick and dead seals began showing up off the coast of Denmark, also killed by distemper. More were found on the shores of Sweden and the Netherlands.[77] The course the disease will take is unknown—and the seals have not recovered from the previous epidemic. But if disease is the problem, where are those sea lion carcasses?

Annual sea lion bycatch kills by Argentine fishing boats range from 170 to 480, mostly young males. That figure is 1–2 percent of the sea lion population in the area and may be about as much as killer whales take.[78] But we don't know what the bigger Spanish, Korean, Taiwanese, and Japanese boats are doing. Most sea lions in the Argentine boat tally are not yearlings but a bit older and, consequently, more "expensive" for population continuance than the very young animals, which have a high risk of dying before breeding age, anyway. Our experience is that the Argentine captains care about what happens to the sea lions. They are cooperative, but no one knows what is happening on the foreign boats. And, there is a large remaining matter.

Between 1935 and 1962, more than 500,000 sea lion skins were taken on the nearby coast of Argentina. Could it be that the Falkland/Malvinas and Patagonian sea lion populations were once the same and the Argentine slaughter thus explains the Falkland/Malvinas disappearance? But if the killing stopped forty years ago, why hasn't the Falkland/Malvinas population, or the Patagonian population, monitored since 1975, made a quick, strong recovery? Although there are scarcely 68,000 southern sea lions on the shores of Patagonia today, the carrying capacity of the area must be, or have been, much greater. In principle, regardless of body mass or population density, a given mass of prey will support a given mass of predator.[79] For example, 10,000 kilograms (22,000 pounds) of prey supports about 90 kilograms (198 pounds) of a given species of land carnivore irrespective of body mass.[80] So there must be less prey or less suitable prey, or some other predator eating the same prey.

Commercial fisheries, like most industries exploiting resources they don't own or have to sustain, are in a mess. The lack of easily enforceable borders in the sea provides all the elements of competition in a "commons." In 1968, ecologist Garrett Hardin pointed out that shared resources, say a pasture where several owners graze cattle on a village commons (or fish in a common fishery), are almost always destined to be overexploited as each user attempts to maximize his take in the absence of strong control. He called the predicament the "Tragedy of the Commons." It has become apparent that sustained successful use of shared natural resources is almost unknown. In fisheries, authorities are dependent on falsified data from fishers and rules that are not enforced. The situation is made even less manageable by climatic uncertainty and political conflicts. Enrique Crespo and his colleagues at Centro Nacional Patagónico suspect that the opportunistic feeding patterns of sea lions protect them, to some extent, from the effects of overfishing. They eat a wide variety of creatures the fishery does not try to catch. Yet competition from the fishery seems the most likely explanation of why the recovery of the sea lion population has been so slow since killing for skins stopped.[81] It is also probable that overfishing has caused ecological changes affecting the basic productivity of the entire coastal system.

The enigma of sea lion disappearance is likely to be joined by a growing number of other sea creature mysteries as populations of invertebrates, such as shrimp, increase, populations of penguins and petrels decrease, and favored fishes disappear. Some will be solved by careful detective work, but it will take a great many conservation cops patrolling the beat to actually prevent them, and a sea change in the management of oceanic resources. As H. L. Mencken observed, "For every complex problem there is a simple answer, and it is wrong."

*When we take the seals from the ocean, what holes are left?*
Victor Scheffer

---

# 2

# *Incredibly Deep-Diving Super Seal*

The pro tem hairdresser jumped back, spilling dark Lady Clairol dye. His "client" responded with a gurgling roar of annoyance. Claudio had inadvertently awakened the big elephant seal while applying letters to her fur that would enable him to recognize her at a distance. Unlike wary sea lions, elephant seals are often approachable. Sleeping, conserving energy, breathing only at intervals, they sometimes seem oblivious of their surroundings. So Claudio and his students were dye-marking and hind-flipper-tagging elefantes by hand, as they began an investigation of their behavior and travels. When the availability of radio satellite tags eventually allowed Claudio to follow his leones at sea, he also looked at Valdés' elefantes. Here were a big sea lion and the world's largest seal living side by side. How were their worlds different, their foods and feeding places? Was the elephant seal population continuing to grow, as it had since Bob Goelet's film crew had counted them in 1964? Would the expanding commercial fisheries threaten their survival? Could their story help win protection for the coast?

The fundamentals of the giant seal's life had been investigated with the smaller northern species in California by Claudio's professor Burney Le Boeuf. Other scientists had investigated the wider-ranging southern seals elsewhere. But little was known about the curiously geographically disconnected Valdés animals. Claudio started by studying them on the beaches. There, the breeding season begins in August with the arrival of the first big bulls. Pregnant females arrive from late August to early October and deliver a single pup about a week after arrival. They care for the pup for a month and wean the youngster by simply abandoning it to feed itself at sea. During the last few days of nursing, the females become sexually receptive and mate several times with one or more males, usually the harem master. By mid-November, all males and females go to sea and forage for about two months, especially for squid. Then they return to the shore to molt, a new stratum of fur pushing out the old outer layers. With a new suit, they go back to sea to forage for six to eight months, put on fat, and prepare themselves for the strain of the next breeding season.

Although the parenting efforts of bull elephant seals are almost nil, their breeding efforts can be prodigious. Claudio named his all-time favorite Odon (aka Odin)—quite a name to live up to:

> Odon was a superb beast: strong, dominant, alert, and a super stud. After following Odon around during the day, carefully observing and recording his behavior, my students and I were exhausted. At the height of the reproductive season, Odon seldom slept, often moved in and out of the harem, chased other males away, herded his females to keep them tightly grouped, and copulated here and there with six to seven females in one hour.
>
> The Don Giovanni of elephant seals, Odon had 122 females at peak season, quite an impressive number for a population in which the average harem size is 20 to 30 females. He was seen mating at least 172 times during the daylight hours with 58 females.[82]

After that breeding season, however, Odon never reappeared. No wonder, you might think. In fact, elephant seal mortality can be very high. For northern elephant seals living along the California and Mexico shore, for example, the mortality rate of females at sea after breeding is about 20 percent; that of males is almost twice as high.[83] Odon's success in sustaining such a huge harem, even if only once, leaves no doubt that he left a major genetic legacy. Many less vigorous males, especially given high mortality rates, never get a chance to mate at all.

Size and color distinguish an elefante pup. Its first pelage is dark brown, which gradually gives way to a soft grayish ivory as it approaches weaning. "Approaches," however, is hardly a hearty enough word. The weight of a newborn is typically ninety to one hundred pounds, and it increases dramatically. Antarctic explorer and scientist Richard Laws, who studied the elefantes of the Falkland/Malvinas Islands, found that weight increased 50 percent in just the first week, doubled at eleven days, trebled at seventeen days, and nearly quadrupled by twenty-one days. Imagine your eight-pound daughter weighing twelve pounds on her seventh day, sixteen pounds on her eleventh day, twenty-four pounds on her seventeenth day, and thirty-two pounds when she is three weeks old. Meanwhile, the nursing mother loses about fifteen pounds a day, eventually 40 percent of her body weight—then she heads for the ocean and food. (Harem bulls may lose more than a ton during the breeding season.) After she leaves, the pup will stay on the beach for up to three months, growing, fasting, and losing about a quarter of its weight, to decline to about three hundred pounds before taking to the sea. Once in the sea, a young male will spend 90 percent of his time submerged—6,000 of the 8,760 hours of the year.

Four out of each ten young will die before the end of the year.[84] By year three, only one in eight northern elephant seal males survives to return to the breeding beaches, and he is still far too young to become part of the reproductive carnival. A Valdés pup may do a bit better; about 38 percent live to be three. Few elephant seal males see their fourteenth birthday, although the females can live to be twenty.

The Valdés elephant seal colony may be the only one in the world that is expanding. Claudio and his CENPAT colleague Mirtha Lewis computed that the population grew at an annual rate of 3.4 percent between 1982 and 2004, only a little slower than the rate of human population growth taking place in less developed countries such as Kenya at the time. Now the Valdés elefantería is almost fifty thousand strong, but there are no other major colonies elsewhere on the coast.

About fourteen thousand females give birth to a similar number of pups grouped in 470 harems in two months. However, a commercial fishery has now developed off the coast of Patagonia and become one of the largest in the world, which may threaten the long-term healthiness of this seal population. So Claudio and his students have begun a study of the Valdés seal's diving and travels, for, ultimately, all animal populations are shaped and limited by their food supply and its location—and so is their conservation.

Mike Hindell and his colleagues at the University of Queensland and Australia's Antarctic Division have recorded amazingly deep dives by southern elephant seals—one of nearly 4,150 feet—and dives that lasted as long as 120 minutes (among humans, the world record is only 6 minutes 41 seconds).[85] Commenting on a dive of 4,125 feet (1,257 meters) off California, Burney Le Boeuf observed, "To put this feat in perspective, imagine three Empire State Buildings stacked one on top of the other. The seal swam from the top of the spire of the uppermost building to the basement of the bottom building in seventeen minutes and then ascended with equal speed to its starting place—a mile-and-a-half vertical round trip in thirty-four minutes. Then, after spending less than two minutes on the surface replenishing its oxygen, it dove to 2,000 feet: then dove again and again."[86] A dive of nearly 5,016 feet (1,529 meters) has also been reported, which, at 14.7 pounds per square inch for every 32.8 feet of descent, means that that seal sustained water pressure of roughly 2,240 pounds per square inch.[87]

Any mammalian diver must overcome two main difficulties: sinking and breathing. The current human record for an unassisted dive on a single breath of air is 236 feet by Umberto Pelizzari of Italy in 1992. The human record for an "assisted" dive (with heavy weights to assist descent and compressed air to jet-propel return to the surface) belongs to Frenchwoman Audrey Mestre, who dove to 558 feet (170 meters) in 2002, before dying three days later in another trial.

The California sea lion has been recorded diving to 904 feet, and a Hooker's sea lion has been recorded at 1,320 feet—all readings obtained by attaching time-depth recorders to the animals.[88] But not much compares with an elephant seal. No other seal can dive as deeply

Andean flamingos feeding at Laguna Lejia, 15,000 feet (4,572 meters), Chile.

Our truck crosses Bolivia's high Kharmapampa on the way to Laguna Colorada.

Southern sea lion male, female, and pup. Peninsula Valdés, Argentina.

Bachelor sea lions raid harems to obtain females, Peninsula Valdés, Argentina.

Andean condor male, distinguished by its fleshy comb, Bronx Zoo, New York.

James' flamingos feeding in the diatom rich waters of Laguna Colorada, Bolivia.

Laguna Colorada reserve, 14,035 feet (4,278 meters) up in the altiplano, Bolivia.

Male Magellanic penguins eye each other suspiciously, Punta Tombo, Argentina.

Magellanic penguins navigate rough seas near Punta Tombo, Argentina.

Graham Harris examines a penguin victim of oil pollution, Chubut, Argentina.

Large bull elephant seal, Peninsula Valdés, Chubut, Argentina.

Claudio Campagna marks female elephant seal with hair dye, Peninsula Valdés.

Elephant seals line Peninsula Valdés shoreline, Argentina.

Veterinarians Robert Cook and William Karesh monitor tranquilized sea lion, Peninsula Valdés.

or as long, and it appears that no whale can dive as deeply or stay down as long—with the probable exception of the sperm whale. Sperm whales may not be immune to the effects of deep diving, however. Recent studies found progressive osteonecrosis on sperm whale bone surfaces, a symptom of decompression sickness, or "the bends." Apparently, no one has looked at elephant seal bones.

Claudio and his colleagues began their investigations of elephant seal swimming and diving with geolocation time-depth recorders (GLTDRs) attached to four adult female seals. These lightweight devices, and the mold of plastic and metal to which they are affixed, are about an inch and a half thick, four inches wide, and seven or eight inches long. Each contains a pressure transducer, a quartz clock, a temperature probe, a light sensor mounted in a clear plastic cap, and a data logger enclosed in a titanium housing. To get diving and swimming data, Claudio had to attach the GLTDR immediately before the seal left on a foraging trip. She would shed the device at her next molt, and with the help of its transmitter, he hoped that he could find her or it on the beach. If she had *not* shed the instrument, recovering it became more interesting, as Claudio related over a few *cervezas* one evening.

Two days before Christmas in 1997, Claudio spotted a tagged female he'd been looking for on the shores of the Peninsula. He was by himself. The female with the radio was in a large group of seals in a dangerously slippery, boulder-strewn portion of shore. How to recover the tag? He debated tranquilizing the seal with his dart gun, but she might head for the water. So, quietly, carefully, he maneuvered himself between the big seal and the water. He had a thin metal towing cable with which to make a catching loop, but it proved too short. So he lengthened it with his shoelaces, eventually lassoed the radio (just as hard as it sounds), and pulled it off the shedding female's loose fur—whereupon it bounced under the snout of a very large and dangerous male. At that point, Claudio was forcibly reminded that the female seals, which surrounded him, could also be dangerous. He was between the "ladies" and the water, and they were displeased.

He found himself trapped behind a large rock by females lunging at him and unable to reach the radio he had bounced under the male's nose. The big fellow finally took a long look at the thing under his chin, and the current disturbance, and galumphed away, but Claudio was still dancing from boulder to boulder, trying to keep away from the teeth of the lunging females. Finally, with a desperate run, leap, and jump, he snatched the radio and, like the other male, galumphed away.

The search for the returning four GLTDR females took place biweekly along a 23-mile stretch of beach. It began in August and continued until December. I well remember Claudio's elation when the gentle-faced Valeria, with a prominent Clairol V on her side, finally returned on Christmas Eve after 70 days at sea. The information recorded, once downloaded into Claudio's computer, showed that she had swum east more than 625 miles (1,006 kilometers) to the vast deeps beyond the continental shelf—where the foreign fishing fleets are hunting. All told, the four GLTDRs on the females recorded 15,836 dives over 270 days. The seals averaged 93 percent of their time underwater and made roughly 2.5 dives each hour, the deepest being Valeria's dive of 3,517 feet (1,072 meters). Particularly important for conservation, however, was the evidence that Valdés seals were not feeding in, or south of, the Antarctic Polar Front. Nor were they hunting in the cold waters along the Antarctic coast, where elephant seals from other colonies whose populations were reported to be declining, or at best stable, were foraging.[89] All of the Valdés seals that have been followed since have shown a similar pattern. And they have also confirmed what every elephant seal diving study notes: The animals spend too little time—an average of 1.6 minutes—at the surface resting and acquiring oxygen to make such lengthy, deep dives possible. It's also been proven impossible for bumblebees to fly.

After the females, seven adult males were tracked between 1994 and 1996, using satellite relay data loggers (SRDLs) attached to their heads. Those devices allowed data to be transmitted when the animals surfaced and their aerials were clear. The transmissions were picked up by the Argos satellite system, which computes the locations from which the transmissions were received and relays them to an Oracle database—thence to Claudio.

The seven monitored seals swam up to 1,300 kilometers (807 miles) from the rookery on Peninsula Valdés, the project revealed; the maximum travel distance recorded for any one seal was 4,500 kilometers (2,812 miles). Unexpectedly, the seals all went in somewhat different directions. Six crossed the Argentine shelf and apparently fished along its deep-sea margin, but one concentrated his activity within a few kilometers of the coastlines, never diving deeper than 94 meters. Most of the dives recorded were along the shelf break at a mean depth of 382 meters (1,253 feet), which would be on the seabed. That phase of the work obtained data from 717 locations and 4,543 dives. Such figures convey a little of the intensity, if not the problems, of some kinds of wildlife research — and they fascinate townspeople and politicians.

All of the dives had been to the seabed, but none of the seals had foraged over the abyssal depths. Thus, the routes of the males had differed markedly from those of the females from the same colony, almost all of which moved beyond the shelf and over the very deep Argentine basin. While Claudio could not with certainty explain why the Valdés elefante population was growing while many other elephant seal populations are in decline, he noted: "The productivity of the Patagonian shelf and the proximity of rich shelf break areas (the largest such shelf in the world) surrounding the colony could play a role in the success of the Peninsula Valdés population. It could provide foraging for young animals with limited diving capabilities and thus increase juvenile survivorship. It is the predictability, stability, and productivity of the southwest Atlantic ecosystem ocean fronts that most likely explain growth during three decades."[90]

The mechanisms by which these divers resolve the conflict between the energetic demands of swimming and the conservation of oxygen stores is not yet well understood. However, time-depth records, velocity profiles, and video images now permit direct observation of swimming periods. Studies on Weddell seal, elephant seal, bottlenose dolphin, and blue whale have revealed that gliding is of major importance in descent, after a short period of stroking. A young elephant seal glided six minutes while descending to 1,300 feet, for example. The glide saves energy and allows the seal to extend its dive time.[91]

To me, the idea of these big, intelligent seals gliding deep within the ocean's immensity far from shore, day after day, months at a time, is a dark and disturbing image. The hidden dangers, the seals' little understood senses, the hunt among the winking bio-lights of big squids and other invertebrates in the lightless depths, the cold, crushing water pressure, are all surreal and shriveling images. The world according to elephant seals is an eerie place. But if the dives are that Herculean, why do them?

The evolutionary stimulus for the elephant seal to develop deep-diving abilities, even beyond those of most whales, apparently lies in its opportunity to tap food supplies available to few other predators. Female northern elephant seals put on about 2 pounds a day while at sea. To do this, they must find and eat about 45 pounds of prey.[92] (To give humans their due, on July 4, 2002, a young Japanese man weighing only 112 pounds won New York's Nathan's hot dog contest by downing an astonishing 50.5 hot dogs with their buns in twelve minutes and gaining seventeen pounds in the process.)

The larder that the elephant seal has found is associated with the "deep scattering layer." In World War II, navy echo-sounders often showed a sea bottom, or several layers, where no bottom existed. The echoes were diffuse and were found to be from strata of euphausids, lantern fish, sergestid shrimps, jellyfish, and even siphonophores. From the elephant seal's point of view, the larger creatures that lunch in these strata, fish and squid, are the target, and they are so far down that the seal is one of their few predators. During the day, depending on latitude, the deep scattering layer stays below 1,800 feet (549 meters), where less than 1 percent of the sun's light penetrates. At night, this layer of organisms swims toward the surface, in part to feed on phytoplankton. Where the light is dim, or at night, when the seals dive as vigorously as during the day, they search for bioluminescent prey, or prey whose movement creates bioluminescent trails. But the main point is that their incredible ability to dive so deeply so often and stay underwater for so long gives them access to a food supply most other air-breathing vertebrates cannot reach.

Seal hunters began exploiting Antarctic seals in the nineteenth century, killing 3,000,000 by 1912. Southern elephant seals were driven nearly to extinction throughout the Southern Cone. In the early 1800s, Spanish, British, and American sealers also began hunting from Mexico and California up the Pacific coast to exploit sea otters and fur seals. They soon discovered thousands of northern elephant seals, and the slaughter began. From the blubber of a single eighteen-foot male elephant seal, a sealer could extract as much as 210 gallons of oil, which was used as lamp fuel. In a few decades, both sea otters and elephant seals were at the edge of extinction. No more than 2,000 sea otters were left worldwide; and, after 1884, when 153 northern elephant seals were killed, subsequent expeditions could find no survivors at all —until 1892, when 8 were found on Guadeloupe Island and 7 of them killed. In 1922, the population had recovered to about 350, and legislative protection in the United States and Mexico was finally given. By 1960, there were said to be 15,000 elephant seals.[93]

When numbers become very small, the obstacles to recovery are so great, the genetic and demographic bottleneck so tight, that few species squeeze through to expand again. Even the most determined conservationists are likely to lose hope. By rights, the northern elephant seal should have been lost, like the Steller's sea cow, the heath hen, and the great auk. Instead, it has made an astonishing recovery—to as many as 180,000. Were there actually many more seals that simply had not been counted?

Working backward, A. R. Hoelzel, from the Department of Genetics at Cambridge, and a distinguished team of geneticists and seal experts, including Claudio and Burney Le Boeuf, undertook a statistical analysis based on the DNA of today's population of northern elephant seals. They also conducted analyses of several southern elephant seal groups, including the Peninsula Valdés colony. With those samples in hand, the team undertook a computer simulation of the northern elephant seal's population growth based on extensive life history information. As expected, they found that northern elephant seals had greatly diminished genetic variability relative to other mammal species and to the southern elephant seal. They concluded that there had been a bottleneck of fewer than thirty seals for as long as twenty years or, if hunting was the primary pressure on the population, a single-year bottleneck of fewer than twenty seals. In fact, the authors observed, "The most likely estimate for a 1-year bottleneck is 10 survivors."[94] Thus, the northern elephant seal offers one of the most extraordinary recoveries from near extinction that we know. It encourages conservationists not to give up. The study also determined that the Valdés population of southern elephant seals might have undergone a reduction itself at some point, although there is no evidence that it was ever large.

As we saw in the steppes, the biomass of a species provides a direct way of gauging the status of a species' food supply (and such pressures as predation and hunting or fishing), whether it be populations of guanacos and rheas indicating the health of steppe vegetation or sea lions and elephant seals revealing the abundance of their foods. Where we have historical information on animal numbers, we can estimate past food abundance.

In his fascinating book *Sobre la Foca Elefante* (*About the Elephant Seal*; 2002), Claudio calculates that the Peninsula Valdés population of 45,000 to 50,000 elephant seals between one and twenty years old weighs about 30,000 metric tons (66,000,000 pounds) and requires approximately 280,000,000,000 kilocalories of energy each year. Those kilocalories have to come from the twenty species of fish and invertebrates, mostly squid, that the seal eats—which produce around 1,000 kilocalories per kilogram. So Peninsula Valdés' elephant seals must consume about 300,000 metric tons (or 660,000,000 pounds), roughly ten times the population's weight, annually. "Is this much or little?" Claudio asks, noting, for comparison, that a human adult requires about 1,000,000 kilocalories each year; thus, the elephant seal's energetic needs are about the same as those of 280,000 humans. "Is it relevant that the world fishery kills and throws away about 29,000,000 metric tons (63,800,000,000 pounds) of fish and marine invertebrates as 'by-catch' each year?" he adds.[95] Such figures surely are relevant to an understanding of the relationships of people and wildlife. It is also relevant that crabeater seals eat 63,000,000 tons of krill each year, and baleen whales about 43,000,000 tons.[96]

If the measures of the elefante's requirements or even the sea's supportive productivity seem so high as to be incredible, consider calculations made by ecologists John Reader and Harvey Croze concerning the biomass of animals on East Africa's Serengeti plains:

> In the Serengeti, the live weight of the 26 species of herbivores, from elephants to dik-diks, is in the region of 1,000,000,000 pounds—roughly the same weight of the inhabitants of Greater London. And that does not include the ten species of primates, 27 species of carnivore, more than 500 species of birds and the uncounted thousands of invertebrate species, which probably weigh as much as all the vertebrates put together. The animal biomass of the wooded grasslands is unequalled in any other living system.[97]

Excluding some coastal and subpolar waters, terrestrial habitats are usually far more productive than the immensely larger oceanic habitats. Although oceans have plenty of sun and water, they are poor in nutrients. About 90 percent of the ocean is "a biological desert," but humans are using close to one-third of the productivity of the other 10 percent through a "catch" that is no longer increasing in size.[98] We are not only reducing the ocean's ability to supply humanity's needs but also leaving less and less for other creatures. This is the ecological theater in which the seabirds and mammals of the Southern Cone must perform. The size of the coast's wildlife community rests on its diversity, its use of different resources in different places at different times. The deep-diving super seal lives a life almost entirely apart from that of the coast's sea lions and penguins and their relatives, but not from its humans.

*Thinking to get at once all the gold the goose could give,*
*he killed it and opened it only to find–nothing.*
Aesop

---

# 3
# *Adventures of a Billion Dollar Bird*

We walked a few cautious steps, stopped, and stood quietly for a while, trying to look like herbivores to avoid frightening the birds. Flights of sharply patterned rock and imperial cormorants sailed in just over our heads, carrying seaweed for nesting. Between six and eight thousand were settled in their traditional rookery atop the rocky plateau at the tip of Punta Tombo, surrounded on three sides by the Atlantic. Big kelp gulls watched us, hoping to use any disturbance to grab an undefended chick. Francisco motioned us up beside him and pointed out the special birds. They were guanays—big colorful Peruvian cormorants, twenty-one hundred miles from home.

Fifty pairs of *Phalacrocorax bougainvillii* had suddenly materialized and colonized Punta Tombo on the wrong side of South America and in the wrong ocean, and Francisco Erize and his colleagues Pablo Canevari and Rodriquez Mata had found them. Francisco is one of Argentina's most distinguished naturalists and wildlife photographers (and has served as president of national parks). Originally lured to Tombo, then very remote, by my work, he had just found the guanays when Bob and I pitched our tents in the distance. He introduced himself in proper fashion, "Goelet and Conway, I presume." This startling guess was delivered with a perfectly straight face and a rather plummy English accent. We were to become close friends.

His second sentence was, "Do you know what a guanay looks like?" Well, yes, we did, but we also knew that no guanay cormorants were closer than the other side of the continent in northern Chile and Peru.

"No. They are here, nesting!" This announcement occurred only after a hurried conference between Francisco and his colleagues. After all, there was the concern of scientific primacy. These Yankees might rush back and publish the discovery before he could. But there was also a niggling worry that these might not be guanays. In the end, he said, "Come and take a look."

Never had there been guanays anywhere on the Atlantic coast of Patagonia. It would be hard to miss them—large black-brown birds with red faces, green eye rings, and white breasts. It was a measure of the unexpectedness of Francisco's find that he wanted confirmation. What made the find doubly fascinating was how fresh the colonization was. Bob and I had been at Tombo in 1964, five years before, and there were none then. Later, Francisco spotted a guanay in a Tombo photo he had taken in 1967. Thus, we can pinpoint the guanay colonization of Tombo as occurring sometime between 1965 and 1967. What were the implications of that colonization, and why then?

Guanays are famous birds. In 1965, Chilean ornithologist A. W. Johnson wrote:

> In strictly commercial terms the Guanay Cormorant is without doubt the most valuable bird in the world as the guano which it produces and deposits on desert islands off the coast of Peru laid the foundation for a world-wide fertilizer industry in the past century and through its extensive use in present day Peruvian agriculture continues to represent a vital asset to that country's economy. Not without reason has it been called the "billion dollar bird."[99]

Johnson did not mention that ownership of the huge amounts of guano accumulated from nesting seabirds was part of the prize over which Peru and Chile fought one of South America's longest and bloodiest wars, finally settled in 1929 after forty-six years. The dispute, originally involving Bolivia, led to the capture of Lima by the militarily superior Chileans, and they held the city for almost three years—until the treaty of Ancón. With the treaty, Peru ceded its most important mainland nitrate province, Tarapaca, to Chile, as well as Tacna and Arica and access to many of the guano birds. At the time, whoever owned the guano bird island deposits had a monopoly on their use as fertilizer. One result was financial collapse for Peru. For many years, much of Peru's federal budget had been paid for with nitrate and bird dung. When the Peruvians recovered the bird islands in 1929, they imposed more losses upon themselves by grossly overharvesting the guano and killing the birds by the hundreds of thousands. The Incas, in contrast, had a highly organized fertilizer industry that carefully protected the producers of the guano by as early as 500 BC.

Every so often, Johnson wrote, guano birds are affected when the cold Humboldt Current deviates from its normal course, coastal waters warm as part of an El Niño, and the guanay's special food, the anchoveta, disappears. The birds disperse, frantically looking for food, "dying by the million and strewing the Peruvian and Chilean coasts with their bodies." Prompted by the sight of more dead and dying birds along the Chilean coast, Johnson brought attention to another issue: "In recent years another serious threat to the Guanay has developed in the form of a large scale fishmeal industry in Peru which competes with the birds for the same source of supply, *Engraulis ringens* [anchoveta]."[100]

Tombo's immigrant guanays had established themselves in the center of the old imperial and rock cormorant colony's main area of perhaps four thousand nests, first recorded at Tombo in 1877.[101] Guanays are highly social, and it is no surprise that they would arrive at Tombo in a group. But why at that particular time? Was it a response to overfishing in Peru, or to an El Niño event? Most likely, it was a combination of the two. During the 1960s, the Peruvian commercial anchoveta fishery expanded almost geometrically. It became so aggressive that the catch made up 15 percent of the world's marine fishery yield, although it was mostly ground up as a fishmeal supplement for livestock feeds. For the guanays, almost wholly dependent on anchoveta, foraging would have become ever more difficult. Then came that 1965 El Niño.

Calculations based on guano production make it possible to estimate the size of Peru's seabird population over time. A guanay weighs about 1.8 kilograms (4 pounds) and deposits about 15.9 kilograms (35 pounds) of guano a year—a little more than that produced by Argentina's imperial cormorant. It is estimated that the seabird population rose from a depressed number of about four million birds, mostly guanays (there were smaller numbers of boobies and pelicans), between 1909 and 1920 to about eight million in the 1930s, before decreasing sharply again in the 1940s. The increase was the product of better protection; the decrease, the result of a series of El Niño events that killed large numbers of birds by making anchoveta unavailable. In the early 1950s, a further recovery led to a population of as many as twenty million birds, almost as many as the estimate of twenty-eight to thirty million for the original population. This was at least partially a result of protecting more of the birds' traditional nesting places. But increase was followed by decline in the late 1950s pursuant to dramatic growth in commercial fishing and then as a consequence of the 1965 El Niño events.

Since that time, the anchoveta fishing fleets have kept the seabird population from fully recovering, even though in 1973, the anchoveta population collapsed and has itself never regained its previous levels. During the past fifteen years, it has become necessary to close the Peruvian fishery from time to time. Predictably, some fishery politicians, ignoring the long association between the huge bird and the huge anchoveta populations, called for a further reduction of the already much reduced seabird population. The facts support no such reduction. If the early-1960s anchoveta stocks were twenty to twenty-five million tons, consumption by seabirds was less than 5 percent. Even the peak bird population of twenty million birds consumed only 6.9–8.6 percent of the stock. More to the point, fish numbers did not increase following the tremendous reduction in the guano bird population, which took place with the 1965 El Niño. All studies show that the breeding seabirds have been relatively minor consumers of fish in the Humboldt Current and also in the comparable fishery in Africa's southeast Atlantic Benguela Current, so culling seabirds is unlikely to improve the landings of these commercial fisheries.[102]

Anchoveta are the Peruvian coast's fundamental economic resource, oxygen to the lungs of the ecosystem. More anchoveta, more human income; fewer anchoveta, less income. Excessive anchoveta fishing to make fishmeal reduces guano birds and, consequently, guano income. It also reduces food for direct human consumption, that food being the anchoveta and the fish that eat anchoveta. Overfishing yields a temporary boon for fishermen but reduces the anchoveta over the long term for fishmeal, for guano, and for wildlife. The relationships are understandable. So far, their management is not.

Chances are that Tombo's venturesome guanays were one of many small flocks of starving cormorants fleeing from both the latest El Niño and the swollen fleets of fishing boats to seek out new sources of food in the mid-1960s. After I saw the Tombo Fifty (pairs) in 1969, I looked for them at every colony I visited along the coast on every visit. In 1972, there were still at least fifty birds (not pairs) at Tombo, and every nest had two or three chicks; in 1974, I counted eighty-four birds. However, Graham and I found only ten active nests in 1981 and none in 1982, although ornithologist Pablo Yorio saw birds there as late as 1994. A few guanays had dispersed from Tombo, whose long-established imperial and rock cormorant colony was shrinking, and appeared at other colonies along the coast of Chubut, mostly mated to imperial cormorants (*P. atriceps*). Several hybrids were produced. The breeding population at Tombo was probably gone by 1986, and now all the Atlantic guanays are gone. For nearly twenty years, we had a rare view of a species trying to spread, though it ultimately failed. To better understand this billion dollar bird, we decided to see it on its home ground.

It was early December of 1995 when Kix, Graham, Claudio, and I arrived in Lima on the way to Punta San Juan, a famous Peruvian guanay, sea lion, and fur seal colony. WCS's Patricia Majluf was our hostess, and we dumped our bags into her Isuzu and the beat-up Blazer of WCS's Peruvian forest expert, Charlie Munn, who was helping.

Patricia, a tall, delightful Peruvian filled with bubbling good humor and local insights, began her scientific work focusing on fur seals at Punta San Juan (PSJ) in 1982. She completed a Ph.D. at Cambridge on their maternal strategies, but her interests had broadened to include all of the Punta's wildlife—the sea lions, Humboldt penguins, guanays, Peruvian boobies, and pelicans—and the overall preservation of Peru's disappearing coastal seabirds and mammals.

Our route to PSJ was a seven-hour drive southwest. The road outside of Lima is usually covered with trash and sometimes with sand. Eventually, it runs through a northern extension of Chile's Atacama Desert, where the Salar de Atacama and its Andean flamingos are found. It ran beside the Pacific beaches and, to our surprise, massive chicken coops on the beaches. Dry, sandy conditions, Patricia explained, help to control disease, and chicken sewage can drain into the sea. We drove through an almost eerie wasteland: huge golden dunes; dusty, desiccated, little roadside hovels; and risky-looking family food stores—where we bought wonderful pickled olives. Signs advertising dolphin meat competed with tiny roadside shrines marking the deaths of loved ones in accidents. Parasails floated over us near Chincha, at the Playa las Toritas, where all drinking water must be delivered by truck. Offshore is the famous Humboldt Current, its cold waters still supporting one of the most important fisheries in the world.

"Eleven million tons of anchovies were harvested along the coast last year but none this year," Patricia observed. "The fishery had to close because of El Niño and overfishing." We passed through the town of Palpa, an important orange growing area, and were saddened by ugly graffiti made of white stones newly inflicted on the nearby hills. They could last thousands of years, as long as the ancient and famous "Nazca Lines" etched into the desert more than a thousand years ago, which Patricia showed us later. After such a kaleidoscope of bizarre and lifeless scenery, Punta San Juan seemed especially wonderful.

Only 54 hectares (133 acres), the rugged Punta is protected from predators and people by a high concrete-block wall built at the suggestion of Robert Cushman Murphy of New York's American Museum of

Natural History. He saw that the birds needed a breeding area free of hunting and predators and that guano was being mined destructively and unsustainably. The government asked him to help in the late 1940s, and he argued successfully for walls and wardens. At the time of our visit, the 54 hectares of PSJ protected about 15,000 sea lions, 3,000–5,000 fur seals, 3,500 Humboldt penguins, 20,000 Peruvian brown pelicans, more than 100,000 guanays, a substantial population of Peruvian boobies, and 47 band-tailed gulls. It was the most complex animal city I have ever seen.

Ignoring the huge, malevolent-looking ticks walking everywhere—gray and about five-eighths of an inch across—I squatted on PSJ's hot, guano-covered ground and watched pelicans landing ungracefully in segregated sections of the colony and long lines of guanays flying back from ocean fishing expeditions.

> When the guanays are moving towards distant feeding grounds they travel not in broad flocks but rather as a solid river of birds, which streams in a sharply marked, unbroken column, close above the waves, until an amazed observer is actually wearied, as a single formation takes four or five hours to pass a given point.[103]

So wrote Robert Murphy, who had the good fortune to visit the Peruvian bird colonies in 1919 before they became so diminished. Despite their greatly reduced numbers, I found the birds' return to their colony at dusk, like Murphy, "the grandest sight of the day." As though the event had been choreographed, long lines of birds flowed single file past high sea rocks, "rising and falling in beautiful undulations" before the deep orange background of a setting sun. Even when nest building, the guanays were coordinated. I took a picture of more than a thousand, each with a feather in its beak held the same way—all at the same time, a music hall fan dance.

Thousands of tons of the hard white guano I sat on was removed in the most recent harvest of this colony before our visit, in 1987, and sold for $200 to $300 per ton—billion dollar birds. Yet, despite the well-known history of mismanagement and its results, despite the wall and the wardens, it had been harvested destructively. No attention was paid to the birds' time of nesting or to the birds themselves. Guano harvesters descended upon PSJ several hundred strong, used digging machines, and killed and ate birds. They defecated all over the colony; frightened off the fur seals, penguins, and sea lions; and created chaos. Patricia has worked for years to gain control of this harvest and finally managed to do so through a 1998 agreement between WCS and PROBONAS, the authority in charge of guano exploitation.[104]

In 2001, Punta San Juan's guano was harvested under Patricia's supervision, over a four-month period by three hundred men from the Andes (*guano* is a Quechua term), the first harvest in thirteen years. It was all dug by hand and in a carefully planned sequence, which confined the diggers to small areas of the colony and left the birds alone everywhere else. Portions of the colony not being harvested were carefully fenced to give the birds security. It worked. Relatively undisturbed, the birds moved back into the harvested areas as soon as the digging was finished (and 13,200 tons of guano obtained). Patricia was ecstatic—and impressed by the responsiveness of the guano contractors: "There was no random cooking or bird killing. They killed a cow every three days to feed their men. Actually, the harvesting system is identical to that of a hundred years ago—except that this time they established proper johns and tents . . . no more human excrement all over the place."

South America isn't the only place where there is fishing conflict over cormorants. North America's double-crested cormorant (*P. auritus*) has become a central figure in another human-bird drama, with some illuminating points in common with fishing conflicts over South America's guanays. Sportsmen and fishers who think that these cormorants, which breed from Newfoundland and Ontario to Cuba, compete with them for fish have declared war. "I hate the ugly things" is the most temperate expression most fishermen of my acquaintance can produce. The situation reveals again the predicament of a fish-eating animal's position in the larger world—as well as the consequences of poor observation and leaps to seemingly unshakeable conclusions. It's reminiscent of the views of some Peruvian fishery politicians.

In North America, double-crested cormorant populations crashed in the 1940s and 1950s, along with those of pelicans, bald eagles, peregrines, and many other birds, following broad-scale applications of DDT and other chlorinated hydrocarbons on agricultural crops and for mosquito control, and also as a result of egging and killing. DDT and its breakdown products alter avian reproductive physiology and, among other things, result in thin-shelled eggs. When broad applications of DDT were finally banned, the then precarious population of the cormorant was finally given protection under the Migratory Bird Act in 1972, and its population began to recuperate. About the same time, fish farming, with its densely packed ponds of growing fishes, became popular, and the recovering cormorants found that there was, after all, a free lunch.

The double-crest's total population in the United States and Canada is now thought to be about two million, and there has been an outburst of claims that it is having a serious impact on both sport and commercial fishing—much noise but little data. According to the American Bird Conservancy, the National Audubon Society, and various other conservation organizations, the fishermen's claims are not supported by any honest science—only by political pressure. The *New York Times* echoed the fishermen's claims with a remarkable article (October 12, 2003), reporting that the cormorant weighs "an average of 3.7 pounds [which is correct], yet they eat roughly 8 pounds of fish a day"—which is ridiculous. Cormorants probably eat only about 5 to 8 ounces a day. Scientists from the distinguished American Ornithologists' Union (AOU) stated, "Every study for about a century has shown that cormorants do not impact significantly the demography of desirable fish, except at very small scales."

The AOU went on to accuse the U.S. Fish and Wildlife Service of giving popular perceptions the same weight as scientific findings by authorizing killing of the "protected" birds. Commercial freshwater fish farmers have been given permission to kill birds damaging their stocks—or looking as though they might. Killing at cormorant nesting islands and elsewhere is widespread. All told, about forty-seven thousand double-crested cormorants are now being killed each year and thousands of their eggs destroyed.[105] This conflict differs from that in Peru, where fishermen claim that guanays and other seabirds are destroying their fishery, in that Peruvian authorities accept the scientific evidence that they are not. The conflict is fundamental and certainly not restricted to cormorants. For example, from 1997 to 2001, the Wildlife Service also authorized the destruction of masses of Caspian terns, great blue herons, black-crowned night herons, and thousands of gull eggs in Washington State because the birds eat salmon.

Puffins, murres, gannets, pelicans, and many other seabirds also eat fish. So do the vast majority of marine mammals, not to mention the completely ignored fact that *fish* eat many times more fish than all the birds and mammals together. For instance, the three million tons of tuna we harvest each year eat about sixty million tons of fish, as we will see later. Moreover, the traditional catch of commercial fisheries has been big, predatory fish such as cod, herring, and mackerel off the U.S. East Coast and hake and robelo off the Patagonian coast, and these are not usually the birds' primary targets. But perceived seabird competition with fisheries, where there is human overfishing, can only intensify. Fishermen do not see fish eating fish.

In contrast to the maligned double-crest in the United States and Canada, some of the Southern Cone's cormorants, as we have seen, are treasured for their nitrogen-rich guano. But guano production in Argentine Patagonia has never been more than a fraction of that in Peru. From 1933 through 1960, it amounted to only 28,154 tons, all told. About 140 tons of that was from the imperial cormorants at Punta Tombo and 10,441 tons from those at Monte León.[106] That harvesting, too, was badly handled, the digging taking place in the middle of the breeding season, with the result that the Monte León colony was reduced to a fragment of its former size and has not yet recovered. Since 1990, there has been little organized guano harvesting in Argentina. Cormorants there are not usually subjected to outright killing, as North America's double-crests are. It probably helps that Patagonia's cormorants are quite beautiful. They are, nevertheless, subject to adverse weather, human disturbance, and predation.

Violent storms and overfishing may have initiated the decline of the historic cormorantry at the tip of Punta Tombo, but the chicks are also very vulnerable to gulls, skuas, and foxes. They are rather pitiful little things, naked, homely, and with long reptilian necks, the kind of creatures only a mother could love; but their mothers do. The gentle way cormorants use their hard, hooked beaks to preen their ill-favored young is quite unexpected. It is also delightful to watch the ways adult rock and imperial cormorants court and snuggle, like college kids in spring—but with the aesthetic advantages of feathers, longer necks, and brighter faces. Cheek to cheek, they sway and bow first on one side, then on the other.

Patagonia's gray cormorants (*P. gaimardi*) build nests on perpendicular cliffs and are almost unapproachable, but imperial cormorants prefer to nest on flat areas and are tolerant of close approach if done slowly and considerately. Too close, and you are warned with deep, guttural growls, spread wings, widely opened beak, and threatening jabs. An unhurried insinuation can win acceptance or, perhaps, just contempt. Imperials build raised, cup-shaped nests like those of guanays, flamingos, and albatrosses in appearance. They are made of seaweed and excreta, the former brought in fresh and dripping from the sea and the latter provided on the spot and augmented by the chicks as they grow and defecate over the edges of the nest. These nests are the main ingredient of the "guano" that is harvested. Rock shags (*P. magellanicus*) build similar nests but, by preference, on steep slopes or in rocky niches in cliffs. Imperials create orderly rookeries, nests arranged precisely, almost geometrically, suggesting domesticity, individual respect, and, less reliably, peace—like guanays, about three per square meter. The drab neotropic cormorant, which nests as far north as the southern United States, nests on bush tops. Of the cormorants left in Patagonia, it is the imperials whose guano has been harvested.

It is remarkable that any species of cormorants are in cold Patagonia, nesting completely around the foot of South America and, except grays and neotropics, even on the Falkland/Malvinas. They appear poorly equipped to deal with cold, as well as to changes in food supply. Studies of the non-Patagonian great cormorants (*P. carbo*), which range through much of the temperate Old World, reveal that they have no more insulation than their relatives from warmer areas, although they spend their winters in Greenland, where the waters dip below 0°C. Patagonia's imperial and rock cormorants have no special cold protection, either, and must maintain their body temperature, like the great cormorant, by what has been termed "intense central thermogenesis," central heating provided by burning a great deal of food in comparison with a well-insulated penguin. But the discovery, by David Grémillet and his colleagues, that great cormorants wintering in Greenland are able to meet their food needs by foraging for an average of only nine minutes a day is nothing less than astonishing. They spend most of their time relaxing, saving energy. They can meet their food needs so quickly because they can catch about a pound of fish in nine minutes.[107] This is much more, as proportion of body weight, than the meals of the maligned double-crested cormorants (or the average meal of a Bengal tiger).

Grémillet concludes that the prerequisite for cormorant success in Greenland is the guaranteed availability of dense fish stocks. The decline of those stocks would probably doom Greenland's cormorants as overfishing and El Niño did Peru's guanays. What does this suggest about the future of Patagonia's colorful cormorants in an ever more heavily fished environment? How hard has it become for a cormorant to feed itself there?

Using VHF transmitters attached to gray cormorants, ecologists Esteban Frere, Flavio Quintana, and Patricia Gandini found that these birds fish by pursuit diving, with most dives made in waters less than fifteen meters deep. When the birds were fishing, they spent 74 percent of their time underwater and an average of less than ten seconds on the surface catching their breath between dives. In contrast to the nine-minute Greenland cormorants, rock cormorant foraging trips lasted nearly two hours, while those of imperials, which fished much farther from the coast, lasted nearly five hours.[108] Patagonia's winter waters may be a little warmer than those of Greenland, but it seems likely that its cormorants are also operating on the physiological edge, and that, no matter where they flee, they are increasingly exposed to the consequences of unrestrained human competition. Although we have no data on how long it took them to find food before commercial fishing began, since the 1960s, the Tombo colony of imperial and rock cormorants has declined from 6,000 to about 50 birds, while the formerly large Monte León colony has still not recovered from guano harvesting long ago.

In hindsight, the cormorant saga suggests that the adventuresome Tombo Fifty guanays never found fish stocks dense enough to sustain their numbers and that the present populations of Patagonia's native cormorants should be carefully monitored. In contrast, North America's double-crested cormorant quickly recovered after DDT regulation; the fish were there.

The big bird in Patagonian coastal waters, in terms of numbers, relations with humans, and biomass, is the Magellanic penguin — the first penguin ever seen by Europeans in the New World. In conservation terms, it is also a seascape species.

*The creatures outside looked from pig to man,*
*and from man to pig, and from pig to man again,*
*but already it was impossible to say which was which.*
George Orwell

---

# 4
# *Penguin Lives and Principles*

Folding my legs between those of my tripod, I scrunched down and tried not to sit on thorns or pointed stones. It was only 5:30 a.m., still too dark for photography, but I consulted with my bladder and decided I was okay for a long wait, warm in an old down jacket, tattered scarf, and some well-worn hiking shoes. The Atlantic was unusually quiet. The tide was low, and the waves were almost three hundred yards away. A Magellanic penguin (*Spheniscus magellanicus*, named for the Straits of Magellan) with two chicks was just twenty-one feet away, according to the scale on the 400-mm lens. I had marked the nest last evening with a handkerchief tied atop a nearby bush, so I could find it in the dark, then slipped out of a bunk in the ranger station early to be here before first light. I was using a 100ASA film and center-weighting the light meter—at least I would when it grew lighter.

The moon was pale and small, and the wind was cold but subdued. It rustled the jume and zampa thickets behind me but lacked enough energy to make a problem of the ever present Patagonian polvo. Penguins sang hoarsely nearby—there were more than 500,000 within a quarter mile of the shoreline. Gradually, sunrise brought life to the land, silhouetting a procession of nine guanacos against wind-thinned clouds and a straw yellow sky. They walked soundlessly atop the barrier beach, single file, their steps raising puffs of that ammonia-acrid mixture of bird dung and fine soil that characterizes giant bird cities in the desert. I was alone in a vast crowd of ten-pound birds, each attending to the business of being a penguin. They were prepared to accept or ignore me so long as I behaved circumspectly. I wanted to know about them, but they did not really care to know about me. I felt as though I was being permitted to view the world of the south Atlantic at its very beginning.

The female penguin, her slightly smaller size, more slender beak, and less coarse face distinguishing her sex, was warming the two chicks her mate had been brooding yesterday. She held them close to her body —one under each wing. I waited to see what kind of sea creatures she would regurgitate to feed the three-week-old nestlings, and to photograph her doing it. She had been foraging at sea about sixty miles from the coast for two days; food-hunting trips were taking that long now, but she looked heavy and full. Now her mate was out there, bobbing in the dawn's dark waves and waiting for enough light to fish on behalf of their chicks in his turn—theirs was an exacting partnership. The chicks begged repeatedly, squeaking incessantly and pushing their soft little beaks at her hard, sharp one.

A black and white kelp gull, circling to land, spotted me and veered away with a stuttered expletive. It had intended to settle beside the penguin family and steal food during the chicks' feeding. A peludo trotted by, her gait amusingly mechanical, stout toenails clicking on the polished pebbles. A distant relative of the giant armadillos and sloths that frequented Patagonia eight thousand years ago, she was intent upon finding something dead or defenseless to eat. I was again at Punta Tombo, 150 miles south of Peninsula Valdés.

Punta Tombo is an isolated, sand-covered, lava-rock peninsula jutting two and a half miles into the south Atlantic. It is at the same latitude as Tasmania, and from the air, it looks like a huge bony finger pointing due east. Twenty-six species of terrestrial birds, mostly small, spend some part of the year here. Larger birds include hawk-like caracaras and chimangos, partridge-sized crested tinamous, and Darwin's rheas. Besides, there are maras, little guinea pig–like cuis, and the *tropilla* (troop) of guanacos. But it is the grand congregation of penguins, perhaps 10 percent of all Magellanic penguins in the world, that makes Punta Tombo so special and qualifies it as one of Earth's great wildlife spectacles.

Tombo is three hours by car from the city of Puerto Madryn at the foot of Peninsula Valdés. You drive the two-lane national Ruta Tres south from Madryn past the outskirts of growing Trelew and over the narrow Chubut River bridge. Then, leaving the blacktop for the gravel of provincial Ruta Uno, you proceed up onto the Meseta de Montemayor, away from people, where paleontologist George Simpson collected huge fossil pythons, ground sloths, and giant penguins. The raised *meseta*, or plateau, is gouged and gullied with dry streams from an ancient ocean, which has drained away, and crisscrossed by low-slung barbed-wire sheep fences. Eighty-one dusty kilometers farther, Ruta Uno turns down toward the sea and continues through the dead village of Dos Pozos to the entrance of La Perla, the big estancia where Punta Tombo is found. From the estancia's entrance to the reserve is twenty-six kilometers.

Descending from the edge of the brush-covered meseta toward the coast, the land becomes increasingly picturesque. Sharp-cornered, brick-red and orange-yellow lava formations protrude through lime-white soil interspersed with waving grasses, gray shrubs, and dark green bushes. Groups of Merino sheep seem to wait hidden in the brush along the edge of the road until you near, then jump in front of your car, the ewes trailing the lambs to be sure to effect maximum consternation. Trios of martinetas hunt for seeds and insects along the road, and the big, antique-looking Darwin's rheas suddenly cross your way, followed by disorderly groups of variably sized chicks, like schoolchildren trailing their teacher on a zoo field trip. And there are always small herds of guanacos, often beautifully posed on the rocks before the distant sea.

The Tombo road eventually leads through a dense grove of evergreens, the infallible clue of human habitation in this treeless land.

They shelter La Perla's estancia house from wind and provide shade. The house is a one-and-a-half-story white stucco and wood structure of about ten rooms with a red metal roof, warm and comfortable in its exotic forest. Over the years, Graham and I have come to know estancia owner Luis LaRegina well, and, when we are lucky, he is home and welcomes us in for conversation and a glass of orange Fanta. He's a big man with iron gray hair, heavy set, florid, with bloodshot blue eyes in a handsome face, red-veined and burnt by seventy years of Patagonian winds and sun; his every word is spoken with conviction.

For the LaRegina family, their historical landlords, Tombo's penguins have become a nuisance—or maybe not. Although the areas where they nest are small, the birds make the land less suitable for sheep, yet the colony is not fenced off from sheep or horses. The estancia's growing guanaco herds have become another nuisance, except as skins. And the tourists, thousands nowadays, complicate the estancia's operation in Luis's view. His son, Alberto, however, observes that the family has built a profitable store and luncheon place adjacent to the ranger station. He enjoys the wildlife and is a defender of the penguins. Luis likes them more than he admits.

Luis was born in 1930. His grandfather arrived from Italy in 1905, spent a year in Buenos Aires Province on a farm near Mar del Plata, then came to Patagonia with Rio Negro sheep shearers, to Cabo Raso, and stayed on when the shearers left. Employed on the Giles Estancia near Camarones, he worked his way up to become manager. Then, hearing of an auction, he bought land (with help from Mr. Giles) and, finally, acquired La Perla in 1927. It covers 17,100 hectares (42,237 acres). Luis and Alberto are filled with information about the land, its history, and its wildlife. I find them enormously sympathetic, although we probably disagree about 30 percent. Graham and I have enjoyed every visit, sitting in the kitchen asking questions, exchanging views, drinking orange soda, and learning—Graham translating most of it for me and, alternately, for Luis.

From the LaRegina house to Punta Tombo is still a considerable distance, but the land grows ever more interesting. Finally, the road tops a hill, and the Punta can just be seen far away, stretching into the blue Atlantic. It is a breathtaking view. From this point, the track flows down through the brush and rocks, winding back and forth like a dry stream to the ranger's residence, the solid white guardafauna station where WCS scientists headquarter. The WCS portion of the building consists of a main room lined with shelves for food, equipment, and data. It is about twelve by twenty feet and has one window, a dining table, some benches and chairs. In one corner, a stove and sink provide cooking facilities; a gas-powered refrigerator occupies part of another corner. Boxes on the floor hold fruits and vegetables, garbage, and trash. Pictures of penguins, maps, and Argentine posters occupy the scant remaining wall space. Behind the main room are a bedroom with bunk beds and a small bathroom with a sometimes functioning shower.

Outside, a Quonset hut provides student sleeping quarters and space for the storage of more equipment. Like the Peludo Palace on Valdés, it is unencumbered by washing and bathroom facilities, but it does have occasional electricity. A trailer behind the Quonset provides a bit more sleeping room, and a nearby storage building for the rangers houses a noisy diesel generator. This simple facility is the Penguin Project's headquarters for several months each year.

Just before the ranger's station, there is a gate, where Tombo entrance fees are collected. After paying, visitors drive past the area where Kix and Bob and I used to tent, then up and over the high barrier beach and down through part of the penguin colony itself to a large, flat area that has been turned into a fenced parking field. This short drive, which is much better walked, provides one of the most spectacular wildlife panoramas anywhere. Although penguins are nesting right up to the ranger station, and sometimes under the trailer, their big colony and the ocean itself are hidden by the barrier beach. Thus, when you arrive atop it, you are totally unprepared for the vast vista of ocean and penguins spread out below you. Many a colleague, inured to spectacular natural panoramas, has been brought to a full stop by this majestic view and the subsequent descent along a road lined with nesting penguins into the heart of the colony. Besides, herds of guanacos and flocks of rheas may be walking along the road in front of you.

# ACT III

Seventy thousand tourists from all over the world now visit Tombo, the number increasing by about three thousand each year. It has become the most popular wildlife attraction in all Patagonia, and its penguin city is both the largest existing Magellanic penguin colony and the most beautifully picturesque. Thousands of penguin nesting burrows cover several thousand acres of bush, rock, sand, and bright Tehuelche pebbles. But the protected fraction of the big colony, the reserve, is a shockingly inadequate 519 acres (210 hectares), and, unlike that of the nearby human towns, its population has declined. Within the area of the reserve, three-strand open-wire fences serve to restrict tourists to broad paths, but not the penguins. The birds walk freely through the tourist areas, and some are so habituated that they nest in the walkways.

From Peninsula Valdés to the Straits of Magellan, there are about 150 coastal colonies of marine birds, several species often nesting together—terns, gulls, cormorants, and penguins. Penguins are the largest, and nearly two million breed along the coast, so they have the greatest biomass of any of these birds (just under ten thousand tons; Chilean, Andean, and James' flamingo populations combined probably have a biomass a bit under five hundred tons). There are also spectacular colonies at San Lorenzo on Peninsula Valdés, Cabo dos Bahias Reserve, Punta Lobería, Punta Medanosa, the beautiful new Monte León national park, and Cabo Virgenes south of Rio Gallegos. This last colony, a large one, is on the very tip of the continent at the mouth of the Straits of Magellan, but it now faces a line of huge offshore oil derricks. As if this uglification were not enough, it is augmented by the deafening and incessant roar of derrick pumps.

At mercifully petroleumless (so far) Tombo, there are nesting populations of deep-voiced imperial cormorants and red-faced rock shags (now, alas, disappearing), of big, flightless steamer ducks, Antarctic skuas, sly dolphin gulls—and thousands of attractive but worrying kelp gulls. There is also a colony of Patagonian sea lions on nearby Islote Chato, and killer whales, right whales, dusky dolphins, elephant seals, big skates, and brown sharks visit regularly.

---

As it grew light enough to take pictures, the female penguin fed her chicks in front of my camera, and I recognized the mantle of a squid and portions of a bony fish in her servings. I wormed close on my belly (involuntarily mopping up penguin dung with my elbows and jacket on the way) and gently pulled a handful of the penguin matron's gooey spillage close enough to recognize bits of anchovy, by far the Magellanic's favorite food. She watched my slow approach and garbage extraction warily, eyeing me with one eye and then the other, but made no complaint.

Crouched in a blind watching terns or lying flat on the shingle observing penguins, I sometimes reflect on the thoughtful travel tales of Paul Theroux. Many of the questions that enter his mind, that extend his perceptions, are subject to up-front testing. He can query the people he is observing and get an answer that is likely to provide insight to their stories, believable or not. When all is said and done, most of natural history and science is telling stories, too, collecting, synthesizing, and putting observations in context. But observation and experiment only indirectly advised by their subjects provide the best material a naturalist can hope for. Penguins keep their thoughts to themselves.

When I backed off and stood up, there were penguins as far as I could see—and hear. Their songs were not birdy tweets but a wonderful cacophony of sonorous, melancholy trumpets. With beak pointed to morning's fading stars or evening's sullen clouds, flipper-like wings held stiffly out from their sides, the vocalists stand as tall as they can and bray their hearts out, Horace's "discordant harmony" personified. The trumpets often end in a series of explosive "huff, huffs," sometimes a threat to rivals infringing on the singer's territory. It is a baritone growl with the sort of vibrancy that I enjoy imagining Caruso offered when he was arrested in the Central Park Zoo in 1906 for pinching a lady's bottom. As midday approaches, Tombo grows quiet.

We don't know whether it was quiet when Magellan passed this way with his small armada in 1520. For that matter, we don't know what it was like when Sir Francis Drake sailed the *Golden Hind* past Tombo in 1578 or when Darwin sailed by in 1833 in the *Beagle*. Nevertheless, from the accounts of Magellan's diarist, Antonio Pigafetta, we suspect that the first penguins to be seen in the New World were spotted somewhere near Punta Tombo.[109] Yet when Henry Durnford actually stopped at Tombo in 1877, he commented on its cormorants but not its penguins.[110] If they were there, how could he overlook them? In any case, they were in force when I visited Tombo in 1964, and my pictures and estimates by other biologists in succeeding years show that the colony was considerably bigger then than it is today. According to Doña Mecha LaRegina, late wife of Luis LaRegina, the birds were there when she was a little girl in the 1930s.

Penguins and other seabirds have repeatedly settled and then abandoned what we now consider traditional nesting areas on the coast of Patagonia. They move in response to food availability, disease, disturbance, and, possibly, predator conditions—a lesson for refuge planners. The first humans to find Tombo probably did not do so until ten thousand years ago, but penguin fossils from nearby date back eighteen to twenty-five million years. They include the giant species that Simpson found, which stood nearly four and a half feet tall and is much older than Punta Tombo itself.

Working closely with Antonio Torrejon, the imaginative and unflagging director of Chubut's budding Dirección de Turismo at the time, WCS began trying to protect Punta Tombo in the late 1960s. Bob Goelet and WCS trustee Nick Griffis provided essential contributions, a tradition carried on by Bob and by the remarkably generous Joan Tweedy for decades.

Progress began when Antonio found a guard to supervise visitors to what we hoped would become a reserve and, at the same time, tried to calm the fears of the LaRegina family. WCS provided a trailer in which the guardafauna could live, and Antonio and I planned a more elaborate ranger station to be jointly paid for by the province and by WCS (with help I was able to get from the Frankfurt Zoological Society). When the permanent building was finally completed, at the west edge of the penguin colony behind the barrier beach, fifteen years had passed since my first visit, but the result was a substantial duplex structure with one side available for the work of WCS scientists. A public restroom was built nearby.

The official opening of the Punta Tombo Reserva Cientifica on December 4, 1979, was more than memorable. The weather was almost balmy, and the ceremony was attended by the governor of the province, a ten-man military band, a kindly black-robed priest, Luis LaRegina, Antonio Torrejon, WCS trustee Frank Larkin, Kix, Graham, and me. Antonio acted as master of ceremonies, the governor spoke, I said a word or two (which no one understood), the military band played (to the appreciative brays of the penguins, which surrounded us), the Argentine flag was raised, and the priest walked cautiously among the penguins sprinkling them with holy water and accompanying his gestures with a blessing or two (which no one understood).

Finally, a protection program was under way and blessed from the start. The infrastructure was in place; Act III was beginning in earnest at Tombo. Now it was necessary to find out what we had let ourselves in for—what it would take to sustain Tombo's penguins and other wildlife. We needed to establish a penguin research and monitoring program, also a training program for Argentines, and help Antonio and other Argentines to develop a local Tombo constituency. We needed an outstanding scientist and special kind of human being.

Dee Boersma, a professor in the Department of Zoology at the University of Washington, had done her dissertation on Galapagos penguins, tropical relatives of the Magellanics. She had won a remarkable string of awards for scholarship and leadership, trained students by the dozen, and produced distinguished scientific reports by the score. It took nearly three years to find her, but she was the right person at the right time, and Tombo proved to be the right place.

Dee is blond, blue-eyed, five-seven, and fit—exceedingly fit. Tough, tireless, enthusiastic, and dedicated are better descriptors. She has now studied at, taught at, and promoted Punta Tombo and its penguins for more than twenty years, and Tombo is not a comfortable place to work. But given the weather, the predators, and the unevenness of the food supply, it is not easy to be a penguin there, either.

It did not take Dee long to find out that, combined with the natural challenges of the Patagonian environment, a growing human population on the coast and an expanding commercial fishery were making Tombo even tougher. When I visited the nearby town of Puerto Madryn in 1964, its population was about 4,500. Today, it is around 100,000, and the even closer towns of Trelew and Rawson have grown nearly as much. How much does the growing human use of coastal resources leave to share with penguins and other wildlife? How much predation can they stand? How much tourism? And how much climatic change, or change in disease prevalence, can the penguin's natural resilience surmount? In brief, what is necessary to secure the survival of penguins at Tombo?

Adult Magellanic penguins usually weigh 3.6 to 5.0 kilograms (8 to 11 pounds) and stand about two feet tall, except when stretching. They wear the usual penguin garb of black and white. Their feathers are small, very dense, and grow atop a layer of fatty blubber that provides exceptional insulation. The birds' beaks are slightly hooked at the tip and can leave nasty scratches on penguin scientists and serious wounds on other penguins. Their legs are short and plump with strong claws, put to use in digging nesting burrows. Although males are bigger than females, the size difference is slightly smaller on average than that between male and female human beings. Inevitably, the upright posture of penguins, not to mention their behavior, invites comparison with humans.

Among Magellanic penguins, as among humans, physical violence is by a wide margin mainly the province of males. The answers to why penguins fight, and they do so fiercely, and why one wins rather than another, are complex and can have serious implications in the Spheniscid world. Daniel Renison, a student from the University of Cordoba, was encouraged by Dee to get some answers. What he discovered, in five hundred hours of observation, was that male Magellanic penguins fight for mates and for nest sites—as expected. But his analysis went further. Where Magellanic penguins find soil that they can excavate, they dig nest burrows as much as three feet deep and six feet long. Where they can't excavate, because the soil is too friable and collapses or too hard for digging, they try to find some sort of shade, most often at the base of a bush. Burrows are the favored housing, though, not just because they provide shade but also because they offer protection from predators and conflicts with other penguins. Thus, nests vary in quality from poor, a simple open scrape in the dirt, to a shady spot under a bush, to the best, a deep burrow with a narrow entrance—a weatherproof place easy to defend from predators. These last are the sites for which penguins fight most vigorously.

The best predictor of winning is size. Bigger birds, like bigger elephant seals, win more fights than smaller ones. Fights for nests can occur after housekeeping is well advanced and even after egg laying has taken place. Under those conditions, the nest owners usually win, even if they are smaller. After all, a nest's benefits to the homeowner with an egg, an incipient family in the nest, are much stronger than the benefits to a stranger. In twenty-two fights before egg laying, size trumped and nest owners won only 55 percent of the time. In seventeen fights after egg laying, owners won 88 percent of the time.[111]

In both sea lions and elephant seals, size advantage enables a big male to fight off smaller males, monopolize the females, and win the reproductive prize of getting his genes into the population. In penguins, monogamy is the rule. A successful bull elephant seal may, in the course of his short reproductive life, sire well over a hundred progeny (remember Odon), though each of his females will rarely have more than seven or eight. In contrast, male and female Magellanic penguins will produce roughly the same number of young in their lifetimes, probably not more than eight or ten. Statistically, no more than two young from any one reproductive pair would be expected to survive to reproduce in a stable population, but few natural populations are stable for long. The survivability of the birds depends on broader forces over which they have no control, as well as their reproductive success.

Important as understanding the behavior of an animal is, knowing its numbers and distribution as well as patterns in the rise or fall of its population is the first step in planning its preservation: How many? Where?

Between 1991 and 1996, all the Magellanic penguin and other seabird colonies on the coast of Argentine Patagonia were censused by air and on foot by Pablo Yorio, Patricia Gandini, Esteban Frere, Graham Harris, and their colleagues. A remarkable atlas was the result.[112] That was how we learned that the coast was inhabited by nearly two million penguins. There are also populations in the Falkland/Malvinas Islands and a few others among the coastal islands of southern Chile. Presently, the Magellanic is one of the few species of penguins—there are seventeen species in all—whose populations are not clearly declining.

Ten species of penguins are diminishing because of some human action, mostly overfishing but also pollution and development at nesting sites. None of these birds is declining because, as some are wont to say, "their time is up," or because "extinction is natural, anyway." Since fishing boats have been appearing off Tombo, Dee has recorded a much less favorable situation at the colony. Some penguins now must go hundreds of miles to find food even while foraging for their chicks. This is a terrible task, given the penguin's modest swimming speed, a maximum of about sixty miles a day. A penguin that has to go a hundred miles to find food for its chicks must spend most of two days just going and coming from the point where it fishes—if it is able to maintain maximum speed.

In the desperate austral summer of 2002, when food was so hard to find that Tombo's penguins were fishing as far as 350 miles away (because of commercial fishing or changes in the density of fish stocks), few eggs were laid and fewer hatched. Few chicks that hatched survived. The usual parenting patterns broke down, and strange things happened in the penguin burrows. Normally, a female lays her eggs and swims off to feed while the male incubates. Then she returns to incubate or help rear the chicks while the male goes to sea to forage, then returns to feed the chicks, and so on. In 2002, in many cases, members of one sex or the other took so long to find food that by the time they returned—if they returned at all—the chicks had starved. Strangest of all, sometimes a bachelor "babysitter male" moved in with a female that had chicks but whose male had disappeared—and the new male would take turns feeding the chicks just as a biological father would.

Adoption in nature, when it occurs, is a rare phenomenon. Each penguin concentrates on perpetuating its own genes, not someone else's. Foster parents occasionally occur in elephant seals, too,[113] but Dee, suspicious, thinks that the babysitter male penguins raising another male's chicks were doing so with "a lively sense of future favors." They were establishing relationships with burrow and female for the next year. Because of their higher mortality from starvation, females were in great demand. Dee believes less in penguin altruism than in planned parenthood. And yet I wonder if there is also a chance that the babysitter is related to the family he adopts—providing an example of kin selection and cooperative breeding. How so? Birds in the immediate area of the nesting female are likely to be birds that were brought up in that area and are related to others there—far-fetched maybe but an alternative to both suspicion and altruism.

Nevertheless, each penguin at Tombo is dependent on its own physical and behavioral tools, abilities bestowed by millions of years of selection in an unremitting struggle to find food, evade enemies, and successfully reproduce in competition with others of its own kind as well as other species. Of course, each bird hones the use of those tools—not their beaks but their behaviors. Eat this, not that; deal with your chicks this way, not that way. Thus a penguin develops personal skills that may ensure its survival and that of its genes. Successfully breeding birds at Punta Tombo, guanacos on the steppes, and flamingos on the altiplano represent an unbroken line of successful parents, grandparents, and great grandparents extending back through time to the founding of life itself. When I listen to the melancholy songs of thousands of penguins as the sun sets at Tombo, I remind myself that it is not an orchestra. Each bird is blowing its own horn.

So Magellanics foraging for fish are hunting selfishly, for themselves and their chicks, alone—but not always. On zodiac trips to some of Patagonia's little-known islands with Flavio Quintana of Centro Nacional Patagónico, Graham and I always come across small *groups* of fishing penguins. One cannot imagine African elephants, except for the big bulls, going off alone to forage—or olive baboons or groove-billed anis. Indeed, studies in Kenya's Amboseli National Park show that the knowledge of older elephant matriarchs is so important to group survival that the loss of an elder female leader results in a reduction in the survival of the herd's babies. Guanacos and many other hoofed animals also feed in herds. But fishers face complications in cooperation unknown to herbivores. So far, the best evidence for collaboration among penguins is in catching schooling fishes.

Rory and Marie-Pierre Wilson found African penguins mostly hunting alone. But when they did fish together, usually ten or fewer birds, they dove synchronously. What happens? They are seeking small fishes such as anchovies, and when they locate a school, they swim rapidly around it, compressing it. Then individuals dive beneath the school and charge through it from below, seizing a single fish, and resume herding the school.[114] The Wilsons calculated that African penguins fishing to feed small chicks travel a mean distance of more than eighteen miles (thirty kilometers) and ingest about 430 grams (15 ounces) of anchovy. Since the average anchovy weighs 7 grams, this led to the conclusion that, when the penguin is in hunting mode, one fish is caught about every five seconds, almost as fast as the rate of a Greenland cormorant.

A Tombo penguin's year might be said to begin at sea off the coast of Rio de Janeiro in early July. To prepare himself for the stresses of defending a nest, incubating, and the other obligations of raising a family, or herself for much the same plus producing eggs, male and female penguins must increase their weights to the maximum. They do this by fishing virtually nonstop for five months, late March to early September. Magellanics can dive regularly below ten meters, and one has even been recorded reaching ninety-one meters.[115] Normally, neither a male nor a female bird will touch land during those months. They hunt and doze at sea in fair weather and foul. They dodge sharks and orcas, perhaps sea lions, and certainly fishing boats. Dee has found that fishing boat nets account for about 6 percent of Tombo's penguin mortality each year, and, until recently, oil spills accounted for 10 percent of their annual deaths (based on hundreds of days spent tramping the shores and examining carcasses by Argentine scientists Patricia Gandini and Esteban Frere). To survive and reproduce, penguins must also escape serious disease and, implicitly, avoid everything humans dump into the ocean that has the potential to affect them.

Each winter and spring, they will swim thousands of zigzag miles foraging yet somehow remember their position with regard to Tombo, for Magellanic penguins are philopatric. They try to return to the same nest or to its immediate area, a considerable navigation feat. At Tombo, some have done so successfully for at least fourteen years and have even been able to identify their burrow among thousands as much as half a mile from shore.

When the time comes, they will read the cryptic maps of waves, sea, and sky, like the South Pacific's legendary Samoan pilots, and, using built-in instrumentation that we do not really understand, head unerringly back to Tombo, but not alone. Not only Tombo's Magellanics and others from the colonies of the Patagonian coast but also Magellanics from the Falkland/Malvinas Islands swim the long, watery wildlife conduit along the edge of the Patagonian shelf on the brink of the Atlantic's abyss. Their route forms a great upside-down Y of hundreds of thousands of penguins swimming north. Each spring, they retrace their journey, separating into regional tribes and migrating back to ancestral colonies on the shores. Much of this ancient seaway has become an area of intense commercial fishing—a frightening gauntlet not only of aqueous furies and age-old predators but also of commercial "long lines" with thousands of hooks, as well as vast murderous nets and uncaring fishermen.

After the breeding season, the penguins swim again to the edge of the Patagonian shelf, where the bottom suddenly drops from one hundred meters to 4,000 meters (13,120 feet). Here they catch the cold, nutrient-rich Falkland Current and ride it north to its confluence with the warm Brazil Current. Reliable food supplies must have been available in that area for a very long time, and the penguins that found them had an advantage over those that did not, so the evolution of an annual journey to use them eventually took place. No doubt, Magellanics have other places that are still secret to find concentrations of food so Dee is following them with satellite tags to see if such centers—sea mounts and ocean fronts, for example—are consistent enough to be identified and set aside as reserves.

Male Magellanics that survive the annual migration south to Tombo will show up on its beaches in late September or early October. Having coped with all the predators, competitors, and other pitfalls that threaten their migration, they now must deal with each other. This means nest selection, defense, and courtship.

Dee summarized penguin courtship about as succinctly as one can:

> Courtship for the Magellanic penguin involves a series of ritual behaviors: bill dueling, mutual preening, circle dancing, and flipper patting; all precede the act of copulation. Copulations are frequent. The male balances precariously on the female's back, treads his feet, and vibrates his bill over the female's bill. With some luck and a lot of cooperation, the vents, the external openings of the birds, meet for an instant and sperm is transferred.

She does not admit that Magellanic courtship is also delightfully amusing. First of all, bill dueling is not just a boy-girl thing. It can take place anywhere, even in the ocean, and most often seems to be a hierarchical behavior between males. Up to eight birds can participate at once. Imagine a football huddle, all the players bent over, helmets close and hands at their sides — banging noses. Because penguin noses are hard, the dueling sounds like the clicking of castanets, as the birds shake their heads from side to side striking each other's bills. But *dueling* isn't quite the right word. While there may be a few deep *ripostes in secunda*, there may also be fierce bites and grabs. Although any encounter can lead to a fight, most often it does not. Whatever they are, duels often play an initiating part in the courtship circle dance.

Now circle dancing, that is very special. Standing tall and stiff, with neck arched, facial feathers fluffed, and eyes half closed, the male circles the female with a mincing gait suitable for a minuet. Finally, he closes snugly in behind, patting her sides rapidly with his flippers. If she is receptive, she bows down, resting her chest on the ground. He mounts her, "balancing precariously," as Dee says, and copulation takes place.

I have spent a fair amount of time eye to eye with penguins wondering how a mated pair recognizes each other among the thousands of look-alikes that surround them. Konrad Lorenz showed that mated geese use facial markings, at least in part. When he altered their faces with a bit of paint, long-mated pairs not only did not recognize but actually attacked each other. Sir Peter Scott's daughter Dafila is said to have been able to recognize the faces of several hundred whooping swans with such accuracy that she could identify birds she had originally observed in wintering flocks in England spotted through a telescope in the Netherlands. It is, when all is said and done, a matter of familiarity and special skill. But colonial birds and mammals are challenging.

It is easy to see that penguins convey emotions and intentions not only with calls and body language but also with their eyes. Some of this eye signaling is not subtle, such as direction of gaze, but other signaling, such as dilation of pupil size between courting penguins, is hard to see. When threatening other penguins—and me—aggressive birds commonly contract their pupils to near pinhole size. They have a great deal of voluntary pupil control. Something similar apparently happens with humans, involuntarily. A recent study has shown that when a woman is "genuinely excited by a man, her pupils tend to dilate, sometimes becoming enormous." In a study in which half the women who were supposedly recruiting men for a psychological survey had their pupils artificially enlarged and half did not, the men volunteered in droves for those with enlarged pupils.[116]

In the 1990s, Dee and her students undertook counts of the entire Punta Tombo colony and determined that there were about 157,000 pairs of nesting Magellanic penguins, a major decline from the 250,000 pairs or more estimated in the 1960s. In each recent year, until 2003–2004, when there was an uptick, their numbers have declined. Here is a section from Dee's 2000–2001 report:

> 2000–2001 was the 19th field season for the Magellanic Penguin Project at Punta Tombo. Adult penguins returned late to the colony, and settlement was protracted: many females first arrived in November and December (usually they arrive in October). Penguins arriving by 25 September were in poor body condition, with the lowest mean settlement weight (4.58 kg) recorded in 18 years, and starved or starving adults were seen on the beaches from September through December.

> Adults also suffered high mortality this season compared to other years. According to preliminary figures, reproductive success was the second lowest (0.052) of the 18 year record, only slightly higher than 1984. More than 65% of nests failed during incubation and about 25% of nests failed because the chicks starved. Failure was due to a lack of food, reflecting poor oceanic conditions for penguins.
>
> On 3 October, 150 dead penguins were found on the beaches in Puerto Madryn. On 9 October, totals included 200 dead adult penguins, as well as more than 400 other seabirds (seagulls, grebes, and cormorants). By 11 October, 1042 dead seabirds had been found in approximately 100 km of beach. Histopathology experts from CENPAT [Centro Nacional Patagónico] hypothesized the die-off was caused by a toxic plankton (possibly demoic acid), however, tissue samples tested did not contain this toxin.
>
> On 1 December, 1072 dead penguins were counted on beaches near Puerto Madryn. The dead birds, mostly adults, were not oiled and did not appear starved. On 5 December, the beaches in Puerto Madryn were covered in dead fish. By 6 December, 3399 dead penguins had been found in approximately 100 km of coastline from Punta Norte, to just south of Punta Tombo. This die-off had been attributed to a toxin from anchovies, but again, tissues tested showed no traces of a toxin.

In happy contrast, the penguins at Punta Tombo did well in 2003–2004, with reproduction the eighth highest in twenty-one years. But only a few years ago, Tombo's penguins numbered *at least* 225,000 pairs, considerably more than the 157,000 pairs estimated at present. Oil pollution, fishery netting, and reductions in food supply probably account for most of the decline. Dee, Graham, and I think there are three reasons for the contraction in penguin food supplies: changes in the numbers and movements of the fishes and other creatures they feed upon (maybe); climate variation (probably); and commercial fishing—fishing boats operating directly within sight of the colony and farther afield and sometimes catching penguins as "bycatch" (certainly). In October 2002, as the birds were trying to come ashore, Alberto LaRegina counted thirty-three fishing boats within sight of the Tombo Reserve, all fishing illegally.

Dee and her collaborators have banded over fifty thousand penguins at Tombo, and they have followed scores of birds on land and sea. From this, it has been possible to calculate not only the colony's overall population trend but also the life histories of individual birds. Of course, the minute a penguin is banded, it is an individual. "Dan" and "Thorne" were two chicks named after Tombo supporter Dan Thorne. "Dan" was banded as a chick in February 1993. Five years later, he became a father for the first time, but both of the chicks he produced with his first mate starved. In 1999 and 2000, the same thing happened.

"Thorne was banded as a chick on February 4, 1989," Dee reported. He was not seen again until 1993; he finally settled down with a mate in 1994 and fledged his first chick. In 1995, predators ate both eggs, but in 1996, the pair fledged two chicks. In 1999, his chicks starved. Dee thinks that the pair spent the winter together at sea because they both arrived at the colony on October 7, 2000. On November 29, his first chick starved, and the second chick was in poor condition. But then, according to Dee, "in late December, conditions improved and by February 14, Thorne and his chick had gone to sea." In 2001, Thorne returned to the same nest for the eighth year and was joined by his mate of five years. Both chicks of that season probably fledged. (In long-lived birds like penguins, the survival rate of adults determines the population's well-being, whereas the health of a population of short-lived birds is usually more strongly influenced by the recruitment rate of youngsters—that is, the addition of youngsters to the population.)[117]

In 1981, just before Dee arrived, the need for a local penguin constituency became unequivocal. A group of Japanese entrepreneurs, styling themselves as the Hinode Penguin Company, joined with several Argentines and made plans to begin "harvesting" from 48,000 to 400,000 penguins at Punta Tombo each year. The birds' skins would be used to make gloves and similar articles, while their meat and fat would be ground up for various animal food products. Some public officials were delighted. After all, that would provide employment for local people (about sixteen to twenty people for three or four months a year). I was flabbergasted. Ignoring the destruction of the penguins, the intent of the recently established Tombo reserve, and the short-term nature of the proposal, the Hinode people had also failed to do their math. None had bothered to calculate how long the approximately 450,000 penguins at Tombo (at that time) would last in such an industry, and, of course, none attached any weight to the intrinsic value of the penguins. Friends in Chubut telephoned frantically, and we quickly produced figures and calculations, sent telegrams, and called authorities. Very soon, we discovered that, since the development of the reserve, the penguins had already begun to acquire a constituency. Hundreds of local people protested and even marched on the governor's house. And that was the end of the Hinode Penguin Company.

By the early 1990s, however, as many as 41,000 penguins were being lost from the Patagonian colonies each year to oil pollution as the oil industry enlarged. Most of the oil came from tankers rinsing their tanks close to shore and from leaky loading areas near Comodoro Rivadavia. As a result of WCS-sponsored studies and the work of Patricia Gandini, Esteban Frere, Dee, and Graham, public concern and the slow but eventually sober response of the oil companies, such pollution is now better controlled.

On the other hand, the need to control commercial fisheries is not being taken seriously by anyone, which is why the valuable hake fishery, the main fishery in the southwest Atlantic in the early 1990s but not a major penguin food species, has steeply declined. Four explanations have been offered for Argentina's failure to maintain a sustainable fishing industry and allow coastal wildlife a better chance to survive. Those explanations are offered in conservation dilemmas from the tropics to the poles. First, perhaps the case for commercial fisheries control has not been persuasively presented to the authorities. But considering what happened to the hake fishery, it is clear that the authorities are not using good fisheries practice in Argentina—any more than they have in the United States or in Europe's North Sea. Second, perhaps the fishing authorities consider that the economic benefits, even from a depleted fishery in coastal wildlife areas, are more important than the persistence of the fishery, the ecotourism the wildlife attracts, or the fundamental preservation of Patagonia's wildlife patrimony. Besides, the people who benefit from tourism (mostly local) and those who benefit from fishing (mostly not local) are different groups, so political conflicts could be a factor. Third, it may simply be that the authorities do not have the resources to control the fisheries and don't say so. Or, fourth and worse, the authorities may be swayed by threats or bribes to ignore the problem. When the guardafauna station opened with provincial and priestly blessings, Punta Tombo's wildlife had only a small constituency. The natural history and the ecological requirements of the birds were largely unknown. Now we know a great deal about Magellanic penguins, and there is a large constituency. Government officials are becoming uneasy about Dee's reports of colony decline, and not only because of the threat to tourism. Local people now take pride in "their" penguins.

Unfortunately, the fisheries dilemma is not "clean." Conservation conflicts almost never are. We cannot be positive that some part of the penguin's food problem is not a result of complex ecological changes, say in currents and the species and populations of the creatures penguins eat. But most probably it is a consequence of overfishing; we will explore that problem later in "The Toothfish, the Hake, and the Squid." When faced with uncertainty, prudent managers of natural resources act conservatively. They restrict fishing to readily sustainable levels and monitor results so as not to damage fundamental assets. They do not hunt sea lions and whales to near extinction or allow sheep to graze pastures beyond the point of recovery.

---

There were no penguins. Waves crashed on the polished pebbles of Tombo's empty beaches, and a few kelp gulls glided by. Patagonian seedsnipes whistled from between the short grass clumps, and a European hare nibbled the stipa, but there were no penguins. Had something terrible happened? There was no hum of crowd noises, chicks mewing, males braying, or vast congregations loafing on the shore. It was deathly quiet. There were no silly waddles, or amusingly serious courtship dances, no clicking of beaks preparatory to a territorial battle, and no wondering tourists.

Every single one of my visits to Punta Tombo has brought me wonder and exhilaration — except one. It was early one August when I was in Argentina to conclude negotiations for the whale sanctuary in Golfo San José. I went to visit Tombo's giant courting, calling, clamoring colony of penguins. None were there. It was as though I had crossed the Brooklyn Bridge to find Manhattan vacant.

August, of course, is penguin winter. The birds were far away, swimming in the high seas to the north. They were fishing, preparing themselves for the stresses of the coming breeding season and a marathon swim of more than a thousand miles back home to Tombo. But their absence contrived a vision of what Punta Tombo would become if ever its penguins were lost. Wonderfully, the austral spring of 2004 provided a different vision, one of blooming health and abundance. The penguins came ashore in exceptional numbers and excellent condition. Even though the long-term population trend is down, they had had a good winter at sea, the best in many years.

Animal colonies such as those at Tombo and the parrot metropolis at El Cóndor have always puzzled biologists. They would seem to make predation and the spread of disease easier and competition for food more severe. Yet in the Southern Cone, colonies are the main nurseries for wildlife over huge areas, especially for seabirds. Their success is critical to conservation in the Southern Cone, and we must try to understand them better. Studies at Chubut's multispecies Punta León colony have produced intriguing findings, so we will go there next.

*Every man is surrounded by a*
*neighborhood of voluntary spies . . .*
Jane Austen

---

# 5
# *The Critical Colony*

Great aggregations of big social animals, cities of penguins, albatrosses, cormorants, and flamingos, the big rookeries of terns, ibises, seals, and sea lions are, despite the Lilliputian scale of their terrestrial sites, the "reproductive generators of wildlife populations over thousands and thousands of square miles," as conservationist Carl Safina puts it.[118] Each is a wildlife superpower, a commanding ecological force in the place where it occurs. Consequently, the survival of these colonial or highly social species is of conservation importance to hundreds of other species as well as to themselves. But if colonies are that important, how do they work?

Punta León is an inconspicuous shoreline bulge below two-hundred-foot cliffs. It consists of a gravel beach, sand, silt, and intertidal rocky platforms, *restingas*, that extend a little way into the Atlantic, and it is a superb example of coloniality. Five species of seabirds nest there in one big colony, cheek by beak, while sea lions breed immediately next door. Difficult access has protected the colony from people so far, but it is less than fifty miles from two growing cities, and there has been much lobbying by local travel agencies to open it to tourism. They forget that cormorants, gulls, terns, and sea lions are not stoics like penguins. These species abandon their colonies when closely approached, and the high cliffs and narrowly confined colonies on the landings below them make it nearly impossible for tourists to see the wildlife without getting close. However, inaccessibility to casual visitors makes Punta León a wonderful place for scientists to work.

Coloniality is usually characterized by tolerance, an absence of the individual feeding and nesting territories that are defended by breeding pairs from others of their kind by animals such as robins, eagles, and pumas. Most bird colonies are associated with water, and 98 percent of seabirds breed in colonies. Because of the vulnerability its concentrations of creatures suggest, however, colonialism is sometimes considered an "evolutionary enigma."[119] Enigma or not, the central purpose colonies serve is reproduction and the defense related to it. Traveling in company means more chances to meet and mate and never having to face a predator alone.

Can you picture a single pair of big, conspicuous, ground-nesting terns, penguins, or flamingos raising its young alone in a world of foxes, skunks, wild cats, hawks, skuas, and gulls? They can't hide their chicks in tree hollows or conceal them on a high branch in a tangle of vegetation, nor can they defend themselves as birds of prey might. Even sea lion pups survive better in colonies.[120] Colonial birds are super predator detectors, and gulls and terns, for whom predation is a constant threat, are effective mobbers and bombers. I speak as a six-foot monster undoubtedly viewed by birds as a "clear and present danger." A bird colony's response to smaller predators, those capable of killing a chick or stealing an egg but not usually of catching an adult, is more direct than mobbing. It consists of masses of jabbing beaks, dramatic threats and screams, aerial attacks, and well-aimed defecations.

The late evolutionary biologist William Hamilton postulated that colonies made sense, selfish sense, for nesting birds. Because a predator would always attack the nearest prey, it would be to the advantage of those potential prey animals that congregate in the open to be close together, as side-by-side nesting pairs would have to deal with only half the "domain of danger" of one alone. If four got together, each would be exposed to predation from only a quarter of the danger's possible directions, and so forth. If you accept this mathematical argument, the larger the colony, the smaller the predation risk to its individual members. But for this explanation, or any other, to be a prime mover in the evolution of colonies, we have to believe that the advantages of colonialism are more compelling than the disadvantages of crowds. Although it is common for small colonies to grow faster than big ones, that seems to occur because larger colonies produce more young than their space allows them to take in and the would-be recruits go to the smaller colonies.[121] As Yogi Berra might put it, "It's too crowded. Nobody goes there anymore."

Nevertheless, members of large colonies must deal with a number of disadvantages: higher competition for food (although group foraging can be an advantage), more exposure to disease, and greater rivalry for nesting space to rear their young. Often, they must defend their eggs, chicks, and nest site even against other colony members; consequently, one parent or the other has to be at the nest until the chick is big enough to defend itself or join a mutually defended crèche. Thus, bird coloniality seems tied to monogamy, although the main reason for monogamy in seabirds may be that male parental care is essential for breeding success, especially for feeding the young. In any event, all colonial seabirds are monogamous. Despite offering relative safety in numbers to inhabitants, colonies provide a supermarket for any capable predators that can reach them, and the constancy of their locations year after year makes them exceptionally vulnerable to human exploitation, as well. The depressing history of the human slaughter of egrets, auks, elephant seals, fur seals, sea turtles, and other colonial animals makes that clear. Nevertheless, it takes only a small benefit over a long period to convey evolutionary advantages to succeeding generations.

That colonies work when human depredations are absent may also have to do with the quality of shared information. Animals acquire a great deal of information about the availability of food by monitoring the actions of others, and there is no better place to do that than in a group. Social cues reveal the location and even the quality of food to many species. By watching, smelling, tasting, and listening, they make use of what ethologists call public information: information that is not purposefully conveyed but is there to be used. The tendency to imitate affects even habitat choice; thus, the formation of colonies, and of breeding and foraging groups.

In most animal colonies, certainly in Patagonia, a further function colonialism serves is to facilitate bringing the sexes together. The influences of massed courtship activity put everyone's hormones on the same schedule and help the process of mate choice, as well as synchronization. Each year along the three-thousand-kilometer coast of Argentine Patagonia, about two and a half million sea mammals and birds participate in an extraordinarily sequential and seemingly coordinated ritual of reproduction—the whales from June to November, the elefantes in September, the penguins from October through January, and the sea lions December through February. This choreography seems especially synchronous in the coordination of breeding.

Initially, we had only a starvation budget to study Punta León's seabirds. However, Graham and Claudio persuaded Pablo Yorio, dark and thin, with incorrigible humor concealed behind a serious demeanor, to take the lead with his student Flavio Quintana, who obtained his doctorate on this project. At first, transportation to the site was provided by Claudio's fourth-hand Peugeot sedan, which lacked a second gear and rarely consented to reverse. The biologists soon realized that they would have to live close to the colony, despite the lack of water and other amenities, and they positioned a tiny WCS trailer atop the Punta's cliffs, with an awe-inspiring view over both ocean and colony.

The bird city is squeezed between the cliffs and the shoreline in an area 600 yards long and 30 to 140 yards wide. The handsome kelp gulls, whose large nesting area surrounds that of the terns and cormorants, occupy the greatest amount of space. Big and aggressive, they probably help to provide protection from ground predators around the colony's perimeter, but they also prey upon smaller colony members.

The researchers found that they could creep and climb through the back of the screaming, cursing, shitting gull community to a flat ledge about seventy-five feet above it. Once there, the birds ignored them, and they could observe the entire theater. To see individual nests close-up, they built a "walking blind." This contraption is made of burlap draped over a folding frame about twenty-eight inches square and six feet high. The observer crawls into this hot, bird-fouled contrivance, bringing binoculars, notebook, and camera; grasps it by its frame; and, with feet protruding incongruously from the bottom, shuffles through the colony directly to the desired observation point. There, the condemned scientist sets the contraption down, kneels within it—peering at birds only a few feet away through holes cut in the burlap—and stays as long as his cramped limbs and excretory system permit.

When I used the walking blind, treading gingerly between nests to find a good vantage point, the birds watched warily but made hardly any protest. Had I tried the same thing without it, they would have taken to the air by the thousands. As I crouched inside, the birds simply detoured around, or perched on top of the blind and defecated through the frayed burlap onto my head and warmly down my neck. Nearby, they carried on with every appearance of normalcy. Royal and cayenne terns hatched before my eyes. Terns carrying tiny fishes landed within inches, bringing food for mates and chicks. A big kelp gull grabbed a robin-sized cayenne tern by the wing, and it was all I could do to keep from breaking cover and separating them—but the gull let go. Beaks pointed high, two graceful cayenne terns renewed their pairing pledges in elegant pirouettes, postures, and chuckling vocalizations. Peering between the errant strands of burlap, I was reminded of Mark Twain's technique for watching a Paris cancan show: "I placed my hands before my face—but looked through my fingers."

Here is the colony census of breeding pairs during one of my visits in the 1990s:

| | |
|---|---|
| Kelp gulls | 9,000 |
| Imperial cormorants | 2,658 |
| Royal terns | 656 |
| Cayenne terns | 1,140 |
| Neotropical cormorants | 96 |
| *Total birds* | *27,100* |

Twenty-seven thousand seabirds flying, courting, arguing, and feeding their young in eleven acres is a panorama both bewildering and fascinating. Extracting useful principles from it takes a person of unusual skill. Pablo Yorio's interest in nature began with a fascination for insects and proceeded through mammals to birds. "As soon as I was old enough to contemplate the idea, I wanted to become a biologist," he told me. But his father sold seeds, agrochemicals, and insurance in Buenos Aires, and Pablo helped him while going to school. Along the way, he became a talented illustrator and cartoonist to support himself. After a series of false starts, he accepted an offer from Dee Boersma of a job on her penguin project. "My life was changed, and that is how my career began," said Pablo. Pablo and his former student Flavio Quintana are now among the most respected ornithologists in Latin America.

Despite Punta León's species mix—two terns, two cormorants, and one gull—Pablo and Flavio found that all the birds start laying eggs at the same time every year, during the second half of October, and that the eggs hatch at nearly the same time.[122] Dee found the same thing with her penguins at Tombo. Synchronous reproduction is key to colony success.

As the bird study continued, it produced data on chick survival and species differences. Just as Hamilton would have predicted, the periphery was less secure than the center. Among imperial cormorants, for example, birds on the periphery were more likely to lose their eggs than those in the center, and fewer of the chicks in their nests survived compared to those nested in the central part of the colony. Among terns, of 454 "predation events" (the great majority carried out by gulls), almost all occurred at the edges of the colonies, not at nests in their center.[123] This peripheralization of predation helps explain some of the advantages of synchronized nesting and of all these species nesting closely together in the first place.

The researchers also discovered that only a few gull families perpetrated the majority of gull predation on tern eggs. I was not surprised. Such individual and independent behavior is common in wolves and jaguars eating a farmer's livestock—and even in visiting herring gulls killing small ducks and pigeons at the Bronx Zoo. Only three individuals, among the sizable flock of herring gulls that regularly visited the zoo, were responsible for the carnage. Since their elimination nearly twenty years ago, no gull attacks on pigeons or teal have been reported. At Punta León's imperial cormorant colony, a single pair of gulls took 24 percent of all the eggs and chicks lost to gulls.[124] At Tombo, I saw two kelp gulls drag imperial cormorant chicks from their nests (on the colony periphery), kill them, attack a chick in another nest, and so on — six chicks killed in less than half an hour by two "perverted" gulls, which, for the most part, did not bother to eat them. The lesson: As wildlife populations become smaller and more vulnerable, control of individual behaviors such as these will become ever more important in conservation.

When the Punta's downy little royal and cayenne tern chicks leave their nests, they gather in dense groups on the beach called, as with penguins, crèches. Moving as one, they resemble a big, fuzzy comforter being pulled over the shingle on hundreds of little black legs — and seem pathetically vulnerable. They are unable to fly, hunt, or feed themselves. Virtually any small predator could kill them. It is not at all clear why the big gulls around do not. Perhaps it is their togetherness, the presence of a few adult terns, or the timing, especially the synchrony, of crèche development in the colony's cycle, for it is a time when the gulls have a great many chicks themselves. In a crèche of seventy-four tern chicks, I counted only four adults. However, tern parents were constantly flying in with small fish to feed their chicks — their own chicks. Unlikely as it seems, chicks and parents recognize each other by sight and sound, as they do in virtually all species of colonial birds, only a few days after hatching. And what occurs at Punta León occurs among colonies all along the coast and in the Falkland/Malvinas Islands.

On Steeple Jason Island in the Falkland/Malvinas, the huge black-browed albatrosses, little rockhopper penguins, and imperial cormorants all had chicks in their vast mixed colony when I saw them in January. Nearby gentoo penguin colonies were filled with larger fledglings, a graduating class of hungry fuzzballs almost as large as their parents. They also kept close together, and the skuas and "Johnny rook" hawks did not attack them—colonial synchrony.

Producing masses of eggs and young all at once in a colony lowers the individual probability of losing a youngster. Ironically, satiating predators by the sacrifice of a few youngsters is the first step in starving them. Suppose you are a fox raising your cubs on the Patagonian coast near a ground-nesting bird colony. Masses of colonial chicks, easy pickings, are available all at once. You gorge, but only for a short time. Soon they fly or swim away as the colony shuts down after its breeding season. Suitable prey living nearby is rare. Scattered cuis, lizards, and such are not likely to sustain your cubs. Where you had been glutted with food, you are now starved. Most of your cubs die. Thus, synchronization minimizes the number of predators awaiting the colony next spring. Synchronization strategies are widespread in nature. Even oak trees use them. Some years, oaks produce only a few acorns, minimizing the number of squirrels that can survive the winter. Then, in a subsequent year, the trees produce a flood of acorns, so many that at least a few are sure to survive the depredations of the minimized squirrel population and get a chance to sprout and become trees. Colony synchronization is as much an inherent conservation effort as a joint defense.

Wherever possible, seabirds colonize areas where ground predators are rare or nonexistent: location, location, location. In addition, almost all Patagonian marine bird or mammal rookeries I know of are a long way from fresh water. They don't need it, and that also makes life difficult for predators that do. Moreover, colonies are usually widely separated, which makes the distances between predation opportunities forbidding. Nowadays, however, these time-tested colonial strategies so essential in the Southern Cone are running into trouble.

Humans are breaking the synchronization-starvation cycle by feeding predators year-round, and that overlooked threat must become a focus for conservation efforts. Humans provide a constant stream of tasty garbage to foxes, gulls, and other predators. For pumas and foxes, we offer savory sheep and European hares, and to buzzard eagles, those hares are perfect fodder, as well. We have even installed windmills and freshwater troughs (meant for sheep) near bird colonies, which make life easier for predators. Thus, seasonal predator starvation and thirst are both being circumvented. The result is that abnormally large numbers of foxes, gulls, and other predators can now survive the annual disappearance of colonialists and attack them in abnormally large numbers on their return to breed—the steppe predator dilemma in different clothing. The margin between survival and death in natural populations is generally so slim that support of the penguins', the flamingos', the cormorants' enemy can only lead to their certain decline. As someone has said, or should have, "The friend of my enemy is my enemy."

Another part of the story is that ancient biological principle "Home is where the heart is." Although some Patagonian terns move their colonies from time to time, that is unusual among colonialists. Just as Dee has found that Magellanic penguins try to nest where they were raised, Claudio has found that sea lion bulls will return to the same stretch of featureless beach where they established their harem the previous year, probably near where they were born. Even those greatest of travelers, the albatrosses, try to breed on the island where they themselves were hatched. More than 90 percent of gray-headed albatrosses nest within 250 meters of where they hatch, and, on Midway, albatrosses nest within 22 meters of the nest where they were born, even though they do not nest at all until they are seven or eight years old.[125] In big vertebrates like these, the charismatic creatures that most often command our interests and sympathies, familiarity with one's surroundings improves chances of finding food, finding safe breeding areas, even of avoiding predation—of survival. Maintaining historical colony sites is thus a high priority for conservation.

A viable population of Magellanic penguins probably requires at least a dozen sizable colonies of several thousand birds each to survive the normal parade of Patagonian catastrophes, now so enlarged by humans. The large numbers of animals immediately apparent in colonies must not mislead us. The traditional wisdom that the probability of a population's extinction is a function of its size is inappropriate to animals that breed in colonies. Because they concentrate in small areas, most colonial species are easily destroyed. Entire populations of congregating species are now dependent on only a few locations for breeding. For them, as we saw for flamingos, *the unit of security is the number of successful colonies rather than the number of animals*. The big colonial creatures that dominate our perceptions of the Southern Cone define their ecosystems in ways that less communal, less synchronized creatures cannot. Their colonies are the nurseries of whole regional populations of their own species and determinants of the populations of many others. If we can ensure that they will survive, so will their worlds.

Besides providing useful insights on the importance of colonial breeding synchrony, and some of the peculiarities of colony structure, the studies at Punta León have proven a successful protection strategy. Pablo, Flavio, and their colleagues developed such a complete picture of the colony that it is now possible to identify adverse effects of exploitation by tourism, so potential users have become cautious. A wild landscape has been defined, responsibility for damage to it can be assigned, and caring people are watching. But the long-term well-being of the coast's ecological giants, its great colonies, is captive to the planning done by the seashore's political authorities and, ultimately, to the well-being of the wildlife in the sea itself.

*I am I plus my surroundings and*
*if I do not preserve the latter, I do not preserve myself.*
José Ortega y Gasset

# 6
# *Plans for Coast and Sea*

PUERTO MADRYN, ARGENTINA. On Friday, March 21, 2004, governors of the provinces of Chubut, Rio Negro, and Tierra del Fuego and the vice-governor of Santa Cruz signed a joint agreement with representatives of the United Nations Development Program and WCS regional coordinator Graham Harris representing Fundación Patagonia Natural. The agreement seeks to implement the Patagonian Coastal Zone Management Plan to protect the biodiversity of Patagonia's 2,000-mile coastal zone, with its spectacular concentrations of wildlife, through "integrated cross-sector participation and research-based decision making." It was endorsed by representatives of Argentina's president Néstor Kirchner.[126]

It was a long time coming. Like a melting ice pack in spring, the agreement was bumped and pushed along by an accumulating flood of information on coastal wildlife and by the public interest it generated. Finally, it broke through. It was created by thousands of days of scientific studies on the long Patagonian shores with their sea lions, elephant seals, penguins, cormorants, and whales, and by countless meetings of fishermen, tourist company officials, provincial authorities, and scientists in every coastal province. But it also came from the impact of growing numbers of tourists attracted by the flood of new insights about Patagonia's coastal animals, which scientists described in newspapers and magazines and on television all over the world.

The plan was developed and led by the Wildlife Conservation Society and Fundación Patagonia Natural, and its implementation is being assisted by the United Nations Development Program and the Global Environmental Facility with a grant of $5.2 million over the coming five years. Where it will go remains to be seen, but this is a major step on the path to a mutually advantageous relationship between coastal wildlife and coastal people. Government departments that up until now worked in virtual isolation will integrate management decisions, with the objective of reducing development pressure on coastal resources in the region, especially fishing. Most important, the agreement lays the foundation for establishing new protected areas in valuable wildlife sites on the coast, as well as strengthening existing sites. A key phrase recognizes its special genesis: "integrated cross-sector participation and research-based decision making." To explain what that means, we must take a step back.

The Patagonian Coastal Zone Management Plan was born in a manuscript for a book chapter. Alfredo Lichter, gifted Buenos Aires scientist and artistic visionary, was planning a book on the marine mammals of Argentina and Antarctica, *Tracks in the Sand, Shadows in the Sea*.[127] Graham would paint its illustrations, and Claudio and other Patagonian scientists would do some of the chapters. Alfredo asked me to write a conservation section, and I responded with "On the Shores of a Cold Sea River," which described the interrelationships of Patagonia's coastal marine mammals, their slaughter, their fragile signs of recovery, and the compelling need for a coastal zone conservation strategy.[128] For reasons I no longer remember, Amy Vedder, WCS's distinguished mountain gorilla ecologist, left my manuscript with an official of the World Bank. Interested, he called Amy and said, "Ask that guy to submit a preliminary proposal by Friday. This looks like something we should do."

Saving biodiversity and its ecosystems is a mandate for the World

Bank's Global Environmental Facility, largely administered by the United Nations Development Program (UNDP). Although the values of the magnificent coastal ecosystem of Patagonia and the threats to it were obvious to me, I well knew how little regarded they were generally; thus, I was surprised as well as delighted when the proposal won encouragement. I called Claudio and Graham, who was struggling to give focus to a newly created conservation organization, Fundación Patagonia Natural, which I had talked him into, and a planning meeting was arranged in Puerto Madryn. It was 1991.

There were nine of us to begin with, nine biologists who saw that the wonders of the Patagonian coast were being lost and could never be saved one by one, province by province. It seemed clear to us, despite the paucity of our knowledge, that it would be necessary to preserve the whole interdependent coastal ecosystem. That, then, was the vision. But how to do it? What would be the strategy? We started by asking ourselves what we thought we knew about the status of Patagonia's coastal wildlife and what it would be like in twenty years if its care continued to be ignored — a typical strategic planning process, except for its subjects and mission. We soon focused on wildlife, fisheries, tourism, and pollution, where we suspected serious and interrelated problems — and realized that we did not know enough, that we lacked in-depth up-to-date information. Only by getting the facts, could we find out what should be done to save coastal wildlife, or convince the authorities to do it — or, for that matter, reason through a strategy that might work. Most of what we did know was a result of WCS's long-time research programs and those of Centro Nacional Patagónico, a national research organization with a branch in Puerto Madryn. But we needed to know much more: We needed to know the kinds and extent of human use of the coastal zone, such as fisheries, tourism, and water contamination. We needed to know wildlife's status and trends. We needed to help strengthen and focus those government and private agencies that exerted authority over wildlife in one way or another. Conservation is not usually an "environmental" problem; it is a human problem. We determined to provide detailed information necessary to both rational development and wildlife protection and push for the creation of an integrated protection program in all the coastal provinces.

Although some of the participatory planning techniques we introduced may have been unfamiliar in Patagonia, their objectives were not, and agreement on what needed doing was unanimous. Subsequently, with help and guidance from UNDP, especially its Latin American officers, Emma Torres and Lita Paparoni, meetings were held from one end of Patagonia's long coast to the other. Each meeting drew from sixty to one hundred participants: government officials, local conservationists, fishermen, tour operators, and scientists. A detailed project proposal resulted, was submitted to UNDP, and eventually garnered $2.7 million, which was increased to over $3 million by WCS.

For the first time, we had enough support to explore the coast's pollution problems, to put observers on fishing boats, to conduct censuses of even the most remote seabird and marine mammal colonies, to provide reports to local legislators and government officials, and, most important, to support focused investigations by an extensive team of specialists. Suave biochemist José Luis Esteves insinuated himself into areas of suspected pollution to obtain samples for his laboratory that his hosts probably wished he wouldn't. Jovial former ranger José Maria Musmeci charmed hoteliers into providing meeting rooms pro bono and restaurants into providing meals for committees at cut-rate prices. Tall, handsome teacher Alicia Tagliorette, whose Spanish is delivered with teletype speed, successfully obtained hitherto unknown facts and figures from tourist agencies and authorities. But it was a slight, red-haired ichthyologist, Guillermo Caille, whose unexpected success most delighted me: getting observers onto Argentine fishing boats. The secret, he explained, is to use women. The (at first) reluctant captains admitted that the result was not just data for the coastal plan but amazing cleanups of both the boats and the language of their crews.

Back in New York at the zoo, we created a geographic information system (GIS) laboratory to prepare coastal maps and identify the areas fieldwork determined to be most important for wildlife and most sensitive to development. Kix and I taped successive drafts of the big map onto our bathroom door and pinned location and species data onto it as Graham relayed them to us, eventually sending our pasteup and GIS specialist to Patagonia to verify details.

The project brought together provincial wildlife officers and scientists from all four coastal provinces to exchange information and work out integrated solutions. One eye-opening investigation after another was conducted, printed, and distributed to local officials—more than fifty, all told. Their subjects ranged from the use of urban garbage dumps by gulls and coastal zone contamination from hydrocarbons, to sea lion trends in northern Patagonia and the effect of tourist disturbance on reproduction, from tourist demands in coastal Patagonian cities to seaweeds of commercial interest along the Patagonian coast.

Over the next three years, my project files grew two penguins tall—Graham's were by now guanaco high. He is the outstanding hero of this extraordinarily difficult, improbable project. It was brought in on time and on budget and is considered a model by UNDP. Scores of officials from both Europe and the United States visited Patagonia to observe its progress—or maybe just to see elephant seals and penguins. An enormous amount of local consciousness raising took place. By the time most of the research and planning were completed, in late 1996, we knew what had to be done and where. The information, and the inclusive, far-ranging process of obtaining it, had also suggested a strategy: integrated cross-sector participation and research-based decision making. First, broadly participatory "coastal committees" would be empowered with funding from the Global Environmental Facility and charged with overseeing improvements in coastal conservation. Second, their decisions would be informed by scientific research. An important part of the process is built-in transparency: public reports and hearings. At last, the development of the coastal plan was followed by the agreement and funding announced in March 2004.

The task facing coastal administrators and committees is to make integrated cross-sector participation and research-based decision making a reality and, while providing recommendations for responsible resource use, to move swiftly to protect the unique creatures of the coast in perpetuity—if there is such a thing. In short, the idea is to treat wildlife as though the people's patrimony depends on it, as it does. As one provincial official is said to have whispered, "This differs from the usual."

One area of growing concern is the very popularity of some of the region's wildlife. From 1997 to 1998, visitors to Chubut's wildlife reserves grew 19 percent, to 279,837 for the year. At Punta Tombo's vast penguin rookery, the best known of Patagonia's wildlife attractions, tourist numbers grew from 55,000 in 1994–95 to more than 70,000 in 2002–03. The studies of Dee Boersma and Pablo Yorio suggest that tourism has not yet adversely affected penguin reproduction at Tombo. Magellanic penguins are stolid citizens. So long as tourists are confined within walkways, which keep them from disturbing nesting concentrations or penguin movement to bring food from the sea to their nestlings, the birds habituate to close association with tourists. However, as mentioned in the section on penguins, the numbers of visitors are now increasing at the rate of about 3,000 each year, and space in the colony for them is not. Thus, the usual visitor is likely to be part of a crowd—an undesirable experience of nature and an obstacle to the colony's interpretation by guides and signs.

Unfortunately, a bus-based attendance pattern has developed, much of it originating from large tour boats that anchor at Puerto Madryn and pay millions of dollars in docking fees. I counted twenty-seven buses at Tombo during a one-hour period in January 2003, a common situation, according to the rangers (57.8 percent of visitors now arrive by bus). Moreover, tour bus schedules usually restrict a Tombo visitor to forty-five minutes with the penguins after four to five hours on the bus. Such a bizarre attendance pattern was previously rare in the world's premier wildlife reserves, but it is becoming more common. More than 300 million visitors crowded into United States parks in 2003. Once intimate lodges at many African reserves have become large hotels. This tells us that people need the same thing the wildlife does, more and larger wildlife reserves. In order to preserve the habitat and make the human experience of the reserves more rewarding, limitations will need to be imposed on the number of people who can visit or stay near especially popular reserves. This is already happening in the United States in places like Yellowstone and Yosemite, but it does not make it easier to protect wildlife.

Even when wild creatures live on privately owned lands, no one really *owns* them. This is especially true of coastal species that are dependent on the sea. Although WCS owns Steeple and Grand Jason Islands in the Falkland/Malvinas Archipelago and Middle Key near Belize as wildlife reserves, long-line fishers and rogue fishing boats are nevertheless destroying the Jasons' magnificent albatross and penguin colonies by their actions at sea thousands of miles away, while fishers and recreational divers are threatening the coral reefs and grouper populations near Middle Key. Protecting terrestrial wildlife on private property might seem easier—but only if there is a way to provide for its security and management, and provided that national and local governance protects sanctuary property rights, and further provided that invasive plants or animals do not appear, or devastating diseases, or poachers, or climate change, or whatever. That is a lot of provisos.

By any measure, Peninsula Valdés, with its guanacos, maras, choiques, sea lions, elephant seals, and more, the "Serengeti of Patagonia," should be a national park, or at least a far better protected provincial park. Development, in the form of increased tourism and demand for more summer mini-estancias, is continuing, gradually debasing this unique part of Patagonia, even though, in 1998, Valdés was recognized as a UNESCO World Heritage site. But the Patagonian Coastal Zone Management Plan will include the entire coast of the peninsula, and, for the first time, local owners and provincial officials have gotten together to address the problem. It is not too late—provided that the effort does not stop on the beach at the ocean's edge.

---

The biggest challenge to the coastal plan's effort to preserve Patagonia's coastal wildlife colonies is carrying its philosophy into the sea itself. Those wildlife concentrations are wholly dependent on the ocean for their food. Consequently, a new effort to preserve the marine wildlife — the fishes and invertebrates, as well as the marine mammals and seabirds, that use the waters over the vast Patagonian continental shelf, extending east to the Falkland/Malvinas Archipelago — is the most fundamental of conservation activities in the Southern Cone. Unlike the steppelands, which are almost all in Argentina, some fifteen jurisdictions affect the southwest Atlantic. But only two are major, those of Argentina and Great Britain (through the Falkland/Malvinas Islands), and the vast majority of the shelf lies within Argentina's Exclusive Economic Zone. Even to attempt to conserve the Patagonian Large Marine Ecosystem seems a bit pretentious, but it is also inspirational in a field sorely lacking in inspiration. Claudio calls the plan now being developed the Sea and Sky Project, if only to suggest its immensity. In terms of wildlife, it might also be called saving the source.

The Patagonian Large Marine Ecosystem, which surrounds and covers the Patagonian continental shelf, the largest in the Southern Hemisphere, is one of the most spectacular and important such natural systems on Earth. Its vast area, two million square kilometers, links Antarctic waters of the south with tropical waters of the north and contains some of the Southern Hemisphere's richest living resources. The whole shelf is 30 percent larger than Alaska and 70 percent the size of Argentina. It supports an enormous animal community, which ranges from sounding whales and soaring albatrosses, to toothfish, squid, and microscopic plankton in a fantastic world of coldly luxuriant seascapes.[129] Animals here are sustained by the nutrient transport system of that cold ocean river, the Falkland Current, by the Brazil Current, and by the frontal zones and upwellings associated with them. Moreover, fully half of the Patagonian marine system is less than one hundred meters deep, and shallow waters are the most productive part of the oceans. Despite the fact that it is the sole source of the food used by the colonies of marine mammals and birds on the coast of Patagonia and the Falkland/Malvinas Archipelago, its management is haphazard, inadequate, and exacerbated by jurisdictional disputes, not least those between Great Britain and Argentina. It is being degraded and overfished, and its wildlife is diminishing.

Marine ecosystems are more open than terrestrial systems. Their ecological barriers are porous and sometimes transient. Such borders and structures as they have, made of currents, temperature, depth, salinity, areas of special productivity, migrations, and fishing pressures, are hard to see, let alone understand. Like terrestrial weather, they change shape and move seasonally, as well as in response to ecological changes, which range from those imposed by the sudden removal of whole populations of fishes or invertebrates by commercial fisheries to those influenced by little understood global cycles. Complicating the situation further are the ephemeral borders of fishing regulations and conventions, and the activities of those who ignore them. The fish and squid fisheries are of global economic importance, and some thirty fishing companies are now managing that huge ecosystem for their own profit and doing it disastrously. In terms of bycatch (the catch of unwanted creatures) and the killing of seabirds, illegal fisheries are a particularly disheartening problem.

As a conservation strategy, the Sea and Sky Project is promoting the implementation of broad ecosystem-based management combined with networks of marine protected areas—again, based on scientific understanding. Its rationale grew out of the data and out of thinking produced by WCS's long-term studies of the region's top predators, the work of Centro Nacional Patagónico, and, especially, the work of the Patagonian Coastal Zone Management Plan. The accumulating reports of decline in commercial fisheries and the tenuous status of the marine mammals and birds led Claudio and Graham to describe the need to face up to the dependency of the coastal wildlife on the sea. With my help, Claudio took the lead, calling upon relationships established in an unlikely meeting between marine mammal and bird experts from the estranged Great Britain and Argentina at the Bronx Zoo in 1995. It developed into a committed partnership between spirited Argentine pinniped expert Claudio and restrained expert John Croxall, of the British Antarctic Survey and BirdLife—despite real distrust generated by the 1982 war. There is no time as good for nongovernmental organizations to act as when governments cannot agree and scientists can.

In 2001, a series of workshops produced an unprecedented pooling of information by thirty-four scientists from Argentina, the United Kingdom (including Falklanders), Germany, New Zealand, South Africa, Australia, and the United States. After meetings with Argentine officials in 2002 and 2003, the project's vision of a truly huge open-ocean protected area plus a network of extractive reserves came together. It would be established on sustainable, science-based, collaborative management and enforced zoning strategies reflecting the variability of oceanic regimes, wildlife populations, and seasonal change. The envisioned strategies include, for example, obligatory onboard vessel monitoring systems, as well as onboard observers and satellite vessel tracking. They also include the development of species-specific bycatch levels and gear of proven effectiveness and, most important, the use of the "precautionary principle" in determining catch quantities—that is, catching less than the maximum levels believed sustainable. In 2004, the concept and supporting data for the plan were placed before a supportive group of expert conservation organizations, including Fundación Vida Silvestre Argentina, the World Wildlife Fund, and the World Conservation Union, in the first of what will be many refinement efforts and the eventual production of a series of recommendations to the nations using the southwest Atlantic.

Ecologically, what is happening in the Patagonian marine system is complex, dynamic, and fascinating. It is, in part, the result of historical exploitation, including whaling and sealing, as well as current fishing practices. Still, current fishing practices constitute by far the largest and most forceful of human impacts. The creatures being fished are enormously important cogs in the machinery of life that supports the ecosystem; they are a necessity for the coastal wildlife, and they are deeply interesting in their own right.

To better see what is happening, we will first visit with the southwest Atlantic's toothfish, hake, and squid. They offer perspectives on human-wildlife relationships very different from those of guanacos and sea lions, as well as newly realized complications. Then, from the detachment of killing cold-blooded fishes and bizarre squid, we will turn to Patagonia's warmly regarded whales and whaling. Finally, we will emerge from the water to consider the albatross—a super, almost supernatural, creature now being watched worldwide, as it provides a life-and-death test of ocean health and wildlife nurseries.

# *Sea and Sky*

*Water and wind, strong currents, high waves, sea and sky: The southwest Atlantic is the restless location of yet another "boundless" area in the Southern Cone. The stormy, richly productive, 2,000,000-square-kilometer (772,200-square-mile) Patagonian Large Marine Ecosystem, the subject of the ambitious Sea and Sky Project, stretches from Argentina to the Falkland/Malvinas and takes in such diverse creatures as squid, whales, albatross, and toothfish.*

*The cod fishery . . . and probably all the great sea fisheries, are inexhaustible; that is to say, that nothing we do seriously affects the number of the fish.*
Thomas Huxley, 1883

---

# I

# *The Toothfish, the Hake, and the Squid*

"Dodging Icebergs, Ships Catch Fish Pirates," reads an August 2003 *New York Times* headline. The pirates in question were illegally fishing for a species of fish that could bring an astonishing $2 million for a single shipload. The recent popularity of the quintessential Southern Cone food fish, the Patagonian toothfish (*Dissostichus eleginoides*), known as Chilean sea bass in Japanese, U.S., and other restaurants, spells danger for the species: It matures so slowly that it is almost certain to be fished out quickly. Its decline, along with the high price it brings, like that of the Australasian orange roughy (*Hoplostethus atlanticus*), has resulted in greatly increased fishing efforts and pirating. The orange roughy has already disappeared from many U.S. restaurants. Less than two feet long, it can live in the wild for more than one hundred years and does not breed until it is between twenty and thirty-two years old. As much as forty thousand tons were being caught each year until the fishery sharply declined. As Tim Flannery says, it is probably not a good idea to "harvest" a creature that lives longer and grows and breeds more slowly than you do. The poor orange roughy was made marketable by changing its name from slimehead, just as "Chilean sea bass" prettified the name Patagonian toothfish.

The main targets of the Patagonian fishery, however, have long been hake, *merluza*, and squid, not toothfish. Analysis of the hake catch led to ridiculed predictions that the fishery would crash, like the cod fishery in the United States had—despite the view of Thomas Huxley—and it did. Catches dropped significantly in 1997 and had not recovered by the end of 2004. Fishing is now controlled—more or less. Meanwhile, squid catches have become unpredictable, varying as much as 1,000 percent from year to year. Such figures suggest that the Southern Cone's marine world is neither stable nor healthy, subject as it is to the double curse of overfishing and illegal fishing.

North American cod fishing supported most of coastal Massachusetts in its early days; it was part of the bounty of the once extraordinarily rich North Atlantic fishing grounds, which have now collapsed in a history that may presage what will happen in the Southern Cone. Cod fishing expanded out from the coasts until it peaked in 1975, after which it sharply declined. As fishing intensity increased eightfold, the biomass of table fish fell by 85 percent. Think of a half-pound Big Mac dropping from 8 ounces to 1.2 ounces. The Eastern Scotia fishery produced dwindling catches of cod until the early 1990s, when fish of usable size vanished. But that was not all. Cod was replaced as top predator of other fishes and invertebrates by hake and seals, and, as a result, more baby cod were being eaten, hampering the recovery of the cod populations. Presently, the new food web along the U.S. northeast coast seems to favor invertebrates such as prawns and lobsters, and their populations have increased; ironically, they are worth more economically, for the time being, than the cod they have replaced.

Patagonian fisheries saw a parallel in the disastrous decline of the hake fishery, which was offset for fishermen by a major increase in—you guessed it—shrimps and prawns. When Graham and I hiked around the perimeter of Punta Tombo in January of 2002, one of the things we found in the big kelp gull and penguin colonies were clumps of disgorged shrimp and red krill (*Munida* sp.) and lots of red penguin droppings, far beyond such leavings in the past. The previous austral spring, of 2001, when penguins were returning to Tombo from their winter migration to breed, there was nothing for them to eat. The breeding areas were ringed by fishing boats that had fished out the area, and penguins began dying on the beaches by the hundreds.

---

John Barton, tall, fit, and laconic, is director of the Falkland/Malvinas fisheries department. At his harborside office in Stanley, he opened a freezer and showed us a big, frozen, exceptionally ugly fish—the first "Chilean sea bass" that Graham, Andrew Taber, or I had ever seen in one piece. "Toothfish," indeed. It had the kind of fang-filled face that nightmares are made of—but it was undeniably spectacular. In fact, marine fishes and invertebrates are among the most spectacular and certainly the most abundant wildlife anywhere in the Southern Cone. They are the foundation, the larder, and the supporting cast of all the coastal seabirds and mammals that live in those violent seas, and their well-being is central to the overall success of much of the Southern Cone's Act III. They have, by far, the region's greatest biodiversity—that almost holy grail of twenty-first-century conservation. If only we could see them more easily.

We can see whales, sea lions, and dolphins when they come up for air or play on the surface, and we see skies filled with seabirds swooping, diving, and living on and shaping that biodiversity, but there are 300 species of fishes in the Mar Argentino, 60 species of bivalves, 105 species of gastropods, and 93 species of crustaceans, all largely unseen. That is more kinds of marine animals alone than all the birds and mammals on the Argentine coast or the Falkland/Malvinas put together. The human exploitation of that unseen multitude, whose biomass is many times that of the creatures we can readily see, is also largely invisible. Of all those creatures, the commercial fisheries capture about 100 species of fishes and invertebrates but actually *land* only 36.[130] Catching or capturing marine animals but not actually bringing them to the dock is not "landing" them. Many captured creatures are unwanted; they are the wrong species or the wrong size, and they are caught, dumped, and called bycatch. Only those 36, and only a portion of them, are considered "the fisheries catch" and are landed. The bycatch is obviously high, and just as dead as the "catch."

In 1997, the Argentine national fishing fleet off Patagonia included approximately 208 vessels. They were mostly trawlers (75 percent), but there were also jigging boats for squid (16 percent) and long-liners (9 percent). Today there are many more, but that national fleet is only part of the marine story of the region. The numbers of boats of foreign origin, from many countries, are also high, all of them fishing year-round in Argentina's "exclusive economic zone." Both foreign and Argentine boats target hake, shortfin squid, and shrimp, but over 25 percent of what they catch has traditionally been species they don't want.

Somewhere between one-fourth and three-fourths of the catch made to supply any seafood dish you eat is thrown away. That is what bycatch means in practice. It is composed of undersized fish and fish of unwanted species, and sea turtles, sea lions, dolphins, and penguins, thousands of petrels and albatrosses, and unwanted invertebrates caught by fishers accidentally. A further 40 to 50 percent by weight is lost as "discharge," which is the garbage that results from fish filleting and processing at sea, according to the Falklands Conservation's Ben Sullivan. What effect is this rain of dead creatures, these spoiling, rotting carcasses, having upon the life and chemistry of the sea? We really have no idea. We do know, however, how devastating the catching process is to albatrosses, petrels, and other seabirds.

Globally, in recent years, it has been calculated that *at least* 27 percent of the total commercial fishery catch is bycatch, 20 to 30 million metric tons of wildlife caught, killed, and dumped each year. In 2004, the UN Food and Agriculture Organization dropped its estimate to 7.3 million tonnes (1 tonne = 1.1 U.S. tons) of wasted fish. This suggests that commercial fisheries may be using more of the fish they used to dump and getting better at avoiding unwanted species. However, it may also indicate just that there are fewer fish to waste. At times, along the Patagonian coast, discards have exceeded 40 percent of the total catch. That is the figure for legal fishing, officially blessed—not illegal fishing. It is in addition to the unreported catch by illegal pirate fishers, boats flying "flags of convenience," and boats fishing in no-fishing zones or out of season or using prohibited fishing gear.

In Patagonian waters, bycatch also may occur after fishers have filled their holds with the species sought but then come across a school of a species selling for more per pound. All too often, they dump the earlier catch and net the latter critters. Then there is the problem of fish size. What are fishers to do with "undersized" hake (*Merluccius*), too small, according to the law meant to protect them so they can grow up and be of catchable dimensions? Fishers must discard them; dead, they become bycatch—fishers can't show up at the dock with them.

In the Southern Cone, the beautiful black and white Commerson's dolphin was regularly killed for crab bait during the 1970s and 1980s—sea lions were, too. In the Pacific, however, tuna fishers have long "set nets on dolphins." This expression means that they watch for dolphins that are chasing fish that are also being chased by schools of foraging tuna, net the whole lot, and kill hundreds of thousands of dolphins each year in order to get the tuna. As a result of public protests, the dolphin bycatch in Pacific tuna fishing by U.S. boats dropped from about 132,000 dolphins killed in 1985 to about 4,000 in 2001. But that is still a lot of big, brainy vertebrates, more than all the black rhinos left on Earth, and almost as many as all the tigers. Besides, most tuna fishermen are not American and still set their nets on dolphins. Albacore (*Thunnus alalunga*), but not yellowfin (*T. albacares*) or bluefin tuna (*T. thynnus*), come as far south as Patagonia, but, as far as I know, nets for those fish are not set on dolphins. Fishers don't set nets on Patagonian sea lions, penguins, or albatrosses, either, yet those animals have become serious bycatch victims—some of the albatross are threatened with virtual extinction.

Fisheries biologist Martín Hall argues that we are not really overexploiting the protein production of the oceans except for the "luxury species" such as bluefin tuna, swordfish, and toothfish. He says, "If we want a large amount of protein, then we should harvest more sardines and anchovies for human consumption rather than for chicken feed." (Almost a third of the marine fishery catch is converted into animal feeds of various kinds.) He continues, "Sardines and anchovies have a biomass orders of magnitude higher than that of species at the top trophic levels"—the big predator fish, that is—just as worm biomass far exceeds that of robins or deer of wolves. He observes that the 3 million metric tons of tuna species harvested each year consume close to sixty million tons of prey. "Yet we choose to harvest the three million metric tons of fancy food rather than the 60 million metric tons of less fancy but still solid nutrition."[131]

Hall's idea is worrying, partly because some say that many populations of sardines and anchovies are already fully exploited, as we saw in "Adventures of a Billion Dollar Bird." Moreover, those little fishes form a fundamental layer in the food web that supports a vast pageant of other species, especially in the southwest Atlantic. The larger challenge is rethinking the global fisheries business. It is out of control and reducing its basic resources as surely as tropical forest timbering and overgrazing on the steppes are reducing theirs.

The seabirds—the albatrosses, petrels, and their relatives—and the cormorants and penguins are especially vulnerable to commercial fisheries. Globally, their populations are going down. Long-liners, gill-netters, and trawlers are the main butchers of albatrosses and petrels, but the trawlers and gill-netters are also catching penguins. Besides, bottom trawling, the practice of dragging nets across the ocean floor is now so widespread and so indiscriminate that it is estimated that every reasonably flat bit of the world's continental shelves is dragged at least once every two years.[132] Such trawling not only scrapes up masses of wildlife the fishers don't want but also destroys the very nature of the seafloor's plant and animal communities. For example, says Juan Kuriger of Ushuaia, the biggest town in Tierra del Fuego, a Japanese commercial fishery destroyed almost all of the Tierra del Fuego crab fishery in just twelve years by using trawls that scooped up everything on the seafloor where the crabs lived and bred.

Edward Melvin and Julia Parrish, biologists at the University of Washington, have addressed the seabird bycatch dilemma of gill-netters in their region with what they consider two tough criteria for success: first, to reduce the bycatch without shifting the major impact to other species and without reduction in the catch of the target species;

second, to find a solution that's practical for fishers and scientifically defensible to managers and conservationists. Their collaboration with gill-netters in Washington's Puget Sound resulted in changing the time of day when the nets would be deployed to avoid the birds' maximum periods of activity, and adding a highly visible panel to the top portion of the net. Those changes helped reduce the seabird bycatch by 75 percent. Similar shifts in practice might have positive effects in the Southern Cone.[133] Other techniques that work with gill-netters might work with long-liners and trawlers in Patagonian waters, too. But is a 75 percent reduction in kill good enough? I wonder about the reasonableness of the Melvin-Parrish criteria. Would we permit such an incidental hunting kill of big nongame land birds? Are we prepared to permit the ongoing killing of already threatened species? Why shouldn't there be a fundamental reduction in the catch of the target species?

Long-line fishing for the slow-breeding toothfish began in the late 1980s. The fish are caught on the seabed at a depth of 800–2,500 meters, mostly near the Falkland/Malvinas and along the edge of the Patagonian shelf. A fisher may deploy more than twenty thousand hooks a day, and it is in hook deployment, or when hooks are being hauled in, that albatrosses, petrels, and other seabirds attracted by the squid bait may be caught and drowned. It happens while the hooks are within three or four meters of the surface. The big albatrosses and petrels do not dive more deeply than that.

If the lines and baited hooks could be deployed so that they sink quickly, bird mortality might be dramatically reduced—even though a single boat's lines may extend for more than fifteen kilometers. Recently, a series of experiments with the two main systems of deploying hooks and lines identified one that could reduce the slaughter of seabirds by southwest Atlantic long-liners rather easily.[134] Unfortunately, new research by Falklands Conservation suggests that, in the Falkland/Malvinas, trawlers are even worse than long-liners in their effects on seabirds and are mostly responsible for the decline of black-browed albatrosses. Basically, however, overfishing is the central marine problem, and those simply trying to "mitigate" the seabird slaughter are not addressing it.

Between 1950 and 1990, world fish landings increased by an astonishing 300 percent. Between 1991 and 2003, the world's industrial fisheries' efforts—though not necessarily their catch—grew by a staggering 22 percent. (In 2000, 95 million tons of wild fish were landed—maybe. The problem is that the biggest supplier of fish, China, has overstated its figures enormously for years. Most experts believe that the catch was much smaller.) Though the size of the global catch seems to have leveled off at 85–95 million tons, nearly 70 percent of the world's fish stocks are already considered fully exploited, overexploited, or severely depleted. High-value fishes, like the Patagonian toothfish, are considered seriously overfished. As the United Nations Food and Agricultural Organization delicately puts it, "Illegal fishing for this species by reflagged vessels appears to be a persistent theme for this fishery."[135] In other words, the toothfish is primarily fished by crooks.

Protected patches of ocean have proven just as useful to marine wildlife as terrestrial reserves have to wildlife on land. However, for many species, those patches have to be very big and even movable to be responsive to seasonal and ecological changes. After all, most marine fish don't live in small areas. They "go with the flow." Hence, the vast zonal areas included in the Patagonian Coastal Zone Management Plan and the Sea and Sky Project. Their care, less visible than that of terrestrial reserves, has to be achieved with *by nots*: by not destroying their nurseries; by not polluting their waters; by not introducing invasive species; by not poisoning them with hydrocarbons, heavy metals, and fertilizers; by not destroying seabed biota by trawling; and by not overfishing. Although the protection offered by broad oceanic *zones* is more responsive to fishing pressures than that offered by fixed *sanctuaries*, reserves can work.[136] A recent review found that population densities of fish reserves, compared to areas outside, averaged 91 percent higher, biomass 192 percent higher, average organism size 31 percent higher, and species diversity 23 percent higher.[137]

Fisheries scientist Ellen Pikitch, whose practical fishing net designs have made a major difference in Alaskan fishery management by releasing undersized fish without killing them, gives a sober assessment of the state of reserves. There are now about thirteen hundred marine protected areas worldwide. "But," she tells me, "half are too small to be effective, there are only forty-one in South American waters, and less than 1 percent have any real protection."

In Argentina, big money from fish, mostly sold overseas, brings in dollars and euros; this stands in contrast to tourism, which is mostly local and therefore brings in only devalued pesos. Dollars also enable some fishing concerns to buy off both regulators and, it is claimed, the press. Elsewhere, illegal fishing has aroused the fishermen themselves. "Some countries, including China, Brazil, South Africa and Taiwan, are infamous for their illegal fishing," says Nelson Beideman, executive director of the Blue Waters Fishermen's Association, representing about one hundred active U.S. long-lining vessels that are calling for multilateral import prohibitions.[138] The United States imports about one-third of the total toothfish harvest.

The good news is, or should be, that the commercial fisheries that affect most of the southwest Atlantic are now overseen by the Commission for the Conservation of Antarctic Marine Living Resources (CCAMLR), called "Camel-lar" by its familiars. CCAMLR has twenty-three members, including Argentina, the United Kingdom, the United States, Brazil, Chile, Japan, the European Union, Russia, and Spain, and it has a constitutional requirement to take an ecological approach to the management of commercial fisheries. The commission has instituted strict fishing and bycatch rules. The bad news is that the rules are not enforced, and it appears that the European Union and Spain are the main protectors of illegal fishing. The rules certainly have not helped the hake, although toothfish fishing may have declined—like the toothfish itself. Recently, single bluefin tunas sold for $157,000 and $173,000 in Japan. Imagine! Neither fish nor fisheries management can survive economics like that. Besides, finding out what's really happening in commercial fisheries is not easy.

It appears that landings from the Argentine commercial fishery more than doubled in the 1990s, then tripled by 2000 to at least 1,200,000 metric tons, and are said to be 1,800,000 tons now—a faster rate of increase than that of commercial fisheries almost anywhere else. About 773,000 tons of that catch came from Patagonian waters. When I first visited the Patagonian shore, the catch was probably less than 100,000 tons. When the all-important hake fishery, which had reached 520,000 tons, crashed in 1997–98, it beggared fishermen and their families all along the coast, but not the fishing companies, which just moved elsewhere. Graham and I photographed abandoned fishing boats along the shores from Comodoro Rivadavia to San Antonio Oeste. For a biomass comparison, there are a total of only about 7,500 tons of sea lions, 10,000 of penguins, and 33,000 of elephant seals presently living along the entire coast—all dwarfed by the biomass of one year's landings of hake.

---

"The merluza are gone, all gone," said Sr. Ojeda when we got past the polite salutations that are a part of any unexpected meeting in Patagonia. We had knocked at the door of his one-room shack on the cold, windy shore near Puerto Santa Cruz. He wore a dusty black boina, and his clothes had been patched and repatched. He looked as thin and worn as they did. "There used to be so many fish, so many," he mused.

Sr. Ojeda has been fishing in this inhospitable place and living alone in this shack for five months each year for more than twenty years. The place had wonderful fishing—it still does. With the merluza (hake) gone, he is fishing for *robalo* (rock cod, *Eleginops maclovinus*) and gets about five hundred kilograms twice a week, which he takes into town to sell. So some fish are still abundant, but not the "luxury" merluza; he gets less for the robalo. He has no electricity or even fresh water, which he must buy in town. His life, like those of many people in Patagonia, is very lonely and very hard. As we leave, apologizing for being unable to share a kindly offered mate, he adds, "There were one hundred fishing boats in the Puerto in January."

---

Seafood consumption per person, at least of the larger species we usually prefer, peaked in 1988. Since then, it has steadily declined worldwide. How could it do otherwise? We have added nearly a billion and a quarter people since then. The fish can't keep up. As commercial fishing has reduced the ocean's top predators, such as swordfish, tuna, hake, and cod, commercial fisheries have shifted toward smaller fishes, the plankton eaters and smaller invertebrates, especially in the Northern Hemisphere. That transition from long-lived fish-eating fish to invertebrates and planktivorous pelagic fish is called "fishing down the food web."[139] A North Sea study calculates that the current biomass of fishes over four kilograms is only about 2.5 percent of its pretrawling level, and the biomass of all fishes is nearly 40 percent lower than it would have been without commercial fisheries. In recent years, as trawlers have lined up off Punta Tombo's shore, illegally, and competition from commercial fisheries has increased, penguins have had to swim farther and farther for food, as noted earlier, and some of their colonies are declining. Areas filled with penguins twenty years ago are now half empty. Dee Boersma calculates that the Tombo colony is down 19.2 percent since 1987. Foraging for his chicks, one satellite-tagged male swam 1,560 kilometers over twenty-seven days, a female 1,428 kilometers in thirty-one days, both without returning to the nest. The chicks can wait about nine days without food. After that, they die. Even with both parents working full-time, the children are doomed when fishing gets that bad.

How can fishery officials calculate reasonable and stable fish catches, optimal "offtakes," for their regulations? Alas, a stabilized ocean fishery producing a constant harvest is an unrealistic dream—unless fishers and fishery managers agree to catches well below those believed maximum, so-called precautionary numbers. Instead, they almost always seek "maximum sustainable yields" (MSYs). These are landings from a calculated "surplus production," which is, ideally, the difference between a fish stock's annual growth and the loss due to natural mortality. The "surplus" idea relies on the fact that all animal species are capable of producing more young than will ever live to reproduce themselves. Thus, removing that "surplus" should not diminish the breeding populations or threaten the survival of the species. Theoretically, a maximum harvestable surplus can be obtained by *reducing* an original animal population by hunting it down to a number thought to be near that of its basic breeding population so that its own density does not limit its numbers through internal competition. After all, most populations include a significant number of nonbreeders. Consequently, fishery theory suggests that the maximum sustainable yield that can be taken from a fishery comes when the biomass of the "target" species is about 50 percent of its original level. However, a great deal must be known about a target population for the theory to work with a margin of safety.

Of course, there really isn't any surplus production in a "stable" wild population in a complex multispecies world, is there? If there were, that population wouldn't remain stable. How can I voice such an apostasy, as my sheepman critic (and a statistically inclined colleague) asked? Sophisticated fishing "take" models have been used for decades. But there really isn't any sure way of separating one species' normal mortality from internal competition of its own kind from that resulting from competition with many other species, diseases, predators, and changing external conditions in a dynamic natural habitat. When we try to maximize the production of just a few elements in a complex system, we change the system. Inevitably, the harvest of a wild species is management under conditions of uncertainty. Although there are smart mathematical models that provide helpful guidelines, smartest of all is the provision of a prudent manager with real authority. He or she must try to track the numbers and size-age condition of the catch while it is actually under way and stop fishers (or hunters) when the caught creatures get too young, the breeding stock becomes at risk, or new factors threaten. At the same time, an eye must be kept on other possibly affected species. There are few "prudent managers" with enforceable authority, however.

John Robinson, far-sighted head of WCS's international conservation program, is expert in the unpleasant realities of tropical forest "harvests." He has compiled a list of maximum percentages of sustainable annual offtakes *derived theoretically from rates of population increase* for some tropical forest mammals.[140] The principles are the same for fish. For example, the maximum percentage of sustainable annual offtake from a healthy population of slow-breeding howler monkeys is about 3 percent; of South American tapirs, about 4 percent; of African duiker antelopes, about 10 percent; and of nine-banded armadillos, about 40 percent. Thus, species with high rates of population increase and normally high mortality should be able to withstand a high offtake. New England white-tailed deer might be able to withstand as much as a 30–50 percent annual offtake. But where reproduction and mortality are naturally low, offtake must be low—as little as 1.6 percent for polar bears, for example. From the little we know of the Patagonian toothfish, it might win a spot on this list of potential exanimates somewhere between tapirs and duikers, but that is not the way it is being fished.

Whatever the math, it is clear that the sea's ability to replenish favorite fishes has been outstripped by the fishing industry's ability to catch them. Global fish catches are declining, while the bycatch kill of seabirds and mammals is increasing because of elevated fishing effort. Seabird bycatch is now well over 300,000 birds each year. They can't stand it. Many are albatrosses and penguins, big slow-breeding animals, slower breeding than most whales and rhinos, and only a little faster than condors or gorillas—certainly as slow as John Robinson's howler monkeys.

Beyond fish, squid (along with octopuses and cuttlefishes, the most wonderfully bizarre of all the mollusks) have become exceptionally important to fishery incomes in the Southern Cone, and it is one fishery whose take has increased, until recently. *Loligo gahi*, a squid whose name sounds like a gaucho sucking the dregs of a mate calabaza, and the more "popular" Argentine short-finned squid (*Illex*) are the beleaguered protagonists.

*Illex* is a very strange creature. It has a one-year life span and is part of the mysterious deep scattering layer, that rich concentration of fish, jellyfish, and squid so important to elephant seals and leatherback turtles but usually to be found down a thousand feet or more in the water column. At night, *Illex* moves toward the surface to feed on zooplankton. It is fished on the southern Patagonian shelf, mostly along the shelf break from just north of Peninsula Valdés south to the Falkland/Malvinas, and is caught by big, surreal-looking "jigger " boats, which fish at night with huge lights. The primary fishers are Japan, China,

Korea, Taiwan, Spain, and Argentina, and the income from fishing licenses for Argentina and the Falkland/Malvinas is huge. Nowadays, the squid harvest accounts for nearly 90 percent of the Falkland/Malvinas' total revenues. In Argentina, the 2002 revenues from the fisheries were 45 percent higher than those from beef. Langostino (shrimp) alone is said to have brought in $400 million. The management challenge in the southwest Atlantic is to keep out illegal fishers, to harvest the fish and squid sustainably, and, for Argentina and the Falkland/Malvinas, to share the stock, whatever it is, on an equitable basis.

I first saw a squid jigger when one anchored off Puerto Madryn. As I scanned the boat with my binoculars, it began a light test. The sudden brilliance of the boat's lights was blinding, like a prolonged flash of lightning. The light comes from rows of stadium-sized incandescent lamps lining the boat's sides, giving it the appearance of a huge octopus tentacle with giant suckers and enough wattage to illuminate a small town. The squid are caught on colored lures armed with a crown of barbless hooks, the "jigs." A single vessel may use as many as twenty-five hundred jigs simultaneously and, under ideal conditions, catch more than thirty tons of squid in one night. Subsequently, the catch is off-loaded to a "reefer," a refrigerator boat, and the jigger goes fishing again.

When NASA recently provided satellite images taken over the south Atlantic at night, we were astonished to discover that among the brightest, most powerfully lit nighttime areas in South America is one located miles off the Patagonian coast. Its lights are nearly as bright as those of Sao Paulo or Rio. It is not Atlantis. It is a gigantic fleet of jiggers using huge lights to attract squid. No one knows how much squid they are taking, or how long that macro ecosystem can last. Until now, the public has been largely ignorant of what is happening, like the stockholders of Enron and WorldCom. All this fishing and killing must be having an effect on the life of the Patagonian shelf, though we don't yet know the details. When these satellite images are coupled with a geographic information system, it is possible to see, for the first time, not only where the jiggers are operating but also, through analysis, how many boats are involved.[141] Thus, to paraphrase Joe Louis, they can run, but they can't hide. We have a new tool for fishery supervision, at least for those boats. Trawlers do not use such lights, and new evidence suggests that the destruction of the sea bottom caused by dragging large nets over it may also be seriously affecting squid populations. Trawlers, not jiggers, catch spawning female squid, which are not affected by lights.

The satellite maps prove that fishing in the Falkland Islands Conservation Zone, the big donut of ocean encircling the Falkland/Malvinas and under its economic control by international agreement, is being stopped by the authorities early enough each year, usually in June, to leave fish and squid breeding stock for the future—as John Barton claims. For example, satellite monitoring revealed no fishing boats in the Falkland Islands Conservation Zone after June 17, 1999, when Barton calculated that more catching might threaten breeding stock. However, it did show major concentrations of fishing boats in the Argentine Exclusive Economic Zone and, especially, on the high seas where the Patagonian shelf break curves outside both Falkland/Malvinas and Argentine zones. Catches in the Falkland/Malvinas zone are monitored so that when squid numbers begin to fall, or, with some fish species, when the catch begins to include undersized fish, fishing is promptly stopped. License fees may be refunded. Nevertheless, the same squid (and fish) populations are being fished by boats operating in Argentine waters and on the high seas, where there is little or no control. This sounds like a prescription for overfishing. The shortfin squid fishery catch did drop dramatically some years ago and then again in 2002 and 2003. But more than overfishing seems at issue. Over one three-year period, the *Illex* catch varied from 260,000 tons to 20,000 tons *with more or less the same fishing effort*. Then, in 2003, it dropped to 3,000 tons. The relationship between squid catch size one year and the size of the stock the next year is unclear. Squid populations appear to be determined mainly by environmental factors, especially water temperature, and, perhaps, by long-term changes not yet understood. There are signs of a global, climate-linked pattern of fluctuation in the numbers of many of the world's largest ocean fish populations, which, in the words of Food and Agriculture Organization experts, "seem to be growing and declining rhythmically, with a high degree of synchrony between them."[142] Failing to take such patterns into account means that the commercial fisheries will hammer stocks during years when they are naturally depressed, making the situation ever worse.

---

What we see in our own lifetimes shapes most of our ideas of what is "normal." With regard to coastal wildlife, that is usually a serious underestimation. There were immense fish populations along the U.S. New England coast when the pilgrims landed. The fish numbers reported are so enormous that people can hardly credit them today. One might say the same thing about the coast of Patagonia, but records are scant and those waters may never have been that productive. I cannot help but think, however, of the huge populations of sea lions killed by sealers there, a half million around Peninsula Valdés alone, and wonder about the profusion of marine life that must have supported them. Where is it now?

In just fifty years, the global spread of industrial-scale fishing has cut the oceans' population of big predatory fishes by 90 percent, from majestic giants like blue marlin, swordfish, and tuna to staples like cod, according to a study by Ransom Myers and Boris Worm of Dalhousie University in Halifax, Nova Scotia.[143] They measured changes in the weight of the big predatory fish, the ones we most like to eat, over time, and found that the biomass of those big fishes is reduced an average of 80 percent within fifteen years of the start of exploitation of a new fishery. Another measure they used was fish caught per hundred hooks on Japanese long lines: In ten to fifteen years, the rate went from ten fish per hundred hooks to only one. New fishing technologies, such as prey-finding sonar and global positioning satellites, have accelerated the decline. Experts believe that recent fish harvests have remained high only because the fisheries have exploited ever more distant fish populations. As ecologist Stuart Pimm notes, you are not going to end up eating diatoms, because they just aren't marketable, but you will eat less seafood, a lot less.[144]

In 2001, Jeremy Jackson and eighteen other scientists collaborated to produce a particularly troubling analysis of the effect of global overfishing, entitled "Historical Overfishing and the Recent Collapse of Coastal Ecosystems."[145] It argues that overfishing of large vertebrates and shellfish has been an extraordinarily far-reaching disturbance to marine ecosystems and that the resulting disappearance of big fishes, marine mammals, and birds has been truly huge and widespread—no surprise in Patagonia. Those animals are now absent from most coastal ecosystems worldwide, and that is one reason Patagonia's remnant coastal fauna is so unusual and fascinating.

There are three major points in the Jackson analysis: "The first is that pollution, eutrophication, physical destruction of habitats, outbreaks of disease, invasions of introduced species, and (even) human-induced climate change all come much later than overfishing in the standard sequence of historical events." The second but most significant part of the analysis contends that overfishing is "a necessary precondition" for most of those unfortunate things to occur. The third point is that "changes in climate are unlikely to be the primary reason for microbial outbreaks and disease."[146]

---

It was summer 2001. Keen-eyed and taciturn, Hormiga, the boatman, cut the Zodiac's motor, and Graham, Flavio Quintana, and I floated quietly in the hot December sun toward the pebble shore of one of the Vernaci Islands in Bahia Bustamante, off the Patagonian shore north of Comodoro. Penguins, kelp gulls, imperial cormorants, and, especially, the endangered Olrog's gull, the subject of Pablo Yorio and Flavio's new study, were nesting nearby. Three big sea lion bulls greeted our arrival with threatening expressions of disapproval. We would have to slink by them as obsequiously as possible. As we landed, Flavio called our attention to plants growing in the clear shallows beneath the boat. "*Undaria*," he said, shaking his head. *Undaria* is an invasive seaweed that has only recently appeared in Patagonian waters, pushing out the native *Gracillaria*. Its structure and its chemistry are different. What will its invasion mean? Will it affect the fishes, the birds? I thought of the seaweed *Caulerpa taxifolia*, recently escaped from an aquarium and now smothering marine life along the shores of California as well as in the Mediterranean.

Offshore, a line of trawlers vacuumed fish from the coastal waters behind Flavio, maintaining their tradition of overfishing. The accelerating traffic of humans, carrying foreign creatures and plants on everything from boats to boots, is ever more rapidly changing the world. We cannot blame climate change for most of our present problems, which include major die-offs of penguins in the Falkland/Malvinas and on the Patagonian coast in 1985 and 2002, although that will come soon enough. The Antarctic Peninsula, reaching out toward South America, has warmed nearly 9°F in the past fifty years, and sea ice has retreated by a fifth since the mid-1970s. Krill populations, the basis of almost all Antarctic wildlife, are in danger of serious decline. Adelie and rockhopper penguin colonies are declining.

Ignoring the fishing boats, we crawled through the island scrub close enough to photograph and count the little colony of elegant Olrog's gulls, the rarest seabird on the entire coast. There were twenty-eight birds.

---

The effects of historical overfishing are synergistic; their overall consequences are greater than the consequences of their parts.[147] It is apparent that they cannot be addressed by anything less than creation of an ecosystem-wide program of fishery management, enforced no-fishing reserves, zones, and bans in the Southern Cone, as proposed in the Sea and Sky Project. There may be no place else that offers such an exceptional opportunity to "do it right." If Argentina and the Falkland/Malvinas, particularly Argentina, with its vast coastal shelf, intelligently manage the southwest Atlantic, they will have a unique opportunity to restore both wildlife and fisheries to historical levels, with long-term benefits for everyone.

Despite the fact that fish are hard to see, fisheries science, especially the techniques of judging the health and sizes of fish stocks, is further advanced than most comparable work on terrestrial animals. Given the number of people involved and the size of the fisheries industry, it certainly should be. It is probably as easy to set prudent population targets for the major "food" fishes as for sea lions, and easier than for guanacos or penguins. The threats to fishes and marine invertebrates, the ways of averting most of those threats and of building up fish populations, are also understood, and the Sea and Sky Project offers a new way to act.

When scientists recommend a no-fishing zone or a ban on fishing for species whose populations are about to collapse, say cod or hake, the industry almost invariably pleads for less drastic regulation and impugns the conclusions of the scientists. Its representatives emphasize the loss of jobs and, sometimes, the "unfairness of considering the fish and not the fisherman." They are defending the employment of people who have followed the fishing trade all their lives, who know no other way to make a living, and who have invested themselves and their families in what has always been a hard and unreliable profession. But such talk ignores survival of the foundation of the fishing life: the fish. When timbering in the U.S. Northwest was restricted as the forests disappeared, there were major protests from those worrying about lost jobs, each argument reminiscent of so many others. The "ivory wars" of fifteen years ago, in which I was deeply involved, are another curious illustration.

In 1988, at a time when African elephants were being killed by ivory poachers for souvenir carvings and piano keys at the rate of one every eight minutes, and the elephant population had dropped 50 percent in ten years, *Time* magazine published a detailed description of the situation, even worse in Asia, and of the elephant's possible extinction. One respondent responded in this indignant way:

> You emphasized the ivory trade of Asia but failed to note the economic impact an ivory ban would have on craftsmen of artistic ivory creations in other parts of the world. Which is more important, an elephant or a human being?[148]

It is hard to understand this kind of myopia. It is so obviously the inchoate expression of one, actually of many, who have never thought ahead to what would happen when the fish, the elephants, or the trees were gone, even in the very near future. Unhappily, the politicians charged with resolving such dilemmas rarely have appropriate training to evaluate them. And fish don't vote.

Is mariculture in Argentina the answer to the need for more fish, to the ecological disasters of overfishing in Patagonia and elsewhere? Not yet. In fact, it appears to be a potential disaster itself. Presently, the rearing and propagation of carnivorous fishes such as salmon and trout require several kilos of ground-up *wild* fish to produce one kilogram of farmed fish. This is provided mostly by catching menhaden, sardines, and anchoveta from wild stocks that in many areas are already being exploited near their maximum sustainable levels. An investigation of Atlantic Salmon of Maine's farms turned up the following information:

> For every pound that a salmon in a farm gains, it devours commercial feed processed from two to five pounds of small open ocean fish like anchoveta, herring and jack mackerel. To satisfy the appetites of the salmon in a one-acre farm, processors vacuum nearly everything bigger than a guppy from 40,000 to 50,000 acres of ocean.[149]

And that is not all of the problem.

Infectious sea lice and salmon anemia have also become ongoing problems in salmon farming. Where sea lice (tiny copepod crustaceans that feed on fish skin and mucous) turn up, millions of fish must be destroyed. And it now seems clear that they are infecting the wild populations—the few that are left. The environmental effects of salmon farming are hard to overstate.[150] But Atlantic salmon farming is now a $2-billion-a-year business producing millions of pounds of fish. "Of the world's roughly 300 million adult Atlantic salmon, only 3.5 million are wild," reports Fen Montaigne in the July 2003 *National Geographic*.[151] About half of the fresh and frozen seafood now consumed by Americans is farmed.[152]

Fish and shrimp farming are expanding because they produce more of selected foods more dependably than fishing and put the profits in the hands of people other than fishermen. They enable the fish farmer to use unrestricted, unmonitored fishes and invertebrates to feed his salmon or shrimps, to control their growth and size, to select the most desirable or easily farmed strains, and to deliver his product more predictably to the market. The fact that such aquaculture is usually destructive to nearby native fish stocks is not his worry. Sea bass and bream are being raised in Italy and Greece, trout in France, and salmon in Great Britain.

Chile is well on the way to becoming the world's leading salmon producer. The escapes of its farmed salmon to the wild and deliberate introductions of farmed brown and rainbow trout in lakes and rivers are certain to affect its forty or more species of native lake fish, of which little is currently known. Preparatory science seems totally lacking.[153] Given the well-known history of tragedy caused by introduced species, such ventures proceeding without due diligence nowadays ought to be punishable by law. The development of fish farms on the coast of Argentine Patagonia to parallel those on the coast of Chile would be a disaster for wildlife.

Marine fisheries, as presently operated, are not nearly as important to the well-being of humans, or as justified, as those fisheries would like us to think. That idea is seditious enough to require numerical justification. How so? To begin with, they provide less than 1 percent of humanity's total food and less than 7 percent of its protein—and those proportions can only decline as our numbers increase.[154] (The total food fish supply reached 94 million tonnes in 1997. However, 33 percent came from aquaculture and 26 percent from freshwater fisheries, leaving only 41 percent from marine fisheries, and as much as 30 percent of that was converted to fishmeal for animal feed.) We get much more of our animal protein from domestic cows, sheep, chickens, goats, and other terrestrial domestics.[155] Moreover, you can be sure that very few poor people are eating the so-called luxury fish. The markets that long-liners and demersal (deep or sea floor) fish trawlers serve are mostly in the wealthy parts of Europe, North America, and Asia. So the fisheries most destructive to marine birds and mammals have few social justifications, even though they are often publicly subsidized. A glaring example is Spain's fishing fleet, a special problem for the wildlife of the Southern Cone. It is Europe's largest fleet, and Vigo, in Galicia, is the largest fishing outfit and port in Europe. The Spanish long ago depleted the fish in their own waters and now send their boats far south to the waters off Mauritania and Patagonia, where they have become gigantic fish sieves. Their operations seem to have scant connection with the strictures of market economics. Not only does the European Union provide about a billion dollars of public monies every year to subsidize EU nation fishing fleets, but about half of the total amount goes to Spain and mostly to Vigo.

In contrast, the artisanal fisheries, the people fishing to feed local communities and families, like Sr. Ojeda in his shoreline hut near Puerto Santa Cruz, fulfill a compelling need—one sometimes threatened by the big commercial boats. For Argentina and the Falkland/Malvinas, the southwest Atlantic fisheries are a gold mine, producing more income than cattle for the former and more than anything else for the latter, through fishing and taxes. If they continue to mine their waters rather than manage them, those waters will become exhausted like gold mines—and the charismatic wildlife will go, too.

So how truly important is seafood, the product of the biggest ongoing slaughter of wildlife on Earth, to us, anyway? For most people, the answer is, not very. For a substantial minority, however, seafood is essential and may constitute as much as 20 percent of their animal protein—but little of that substantial minority provides a primary market for the kind of fishing that includes major bycatch of slow-breeding seabirds and mammals. At some point, commercial fisheries must agree on what constitutes the *optimum*, not the *maximum*, catch of swordfish, hake, anchoveta, toothfish, or whatever. Those who determine that figure *must* then insist that bycatch of slow-breeding species be eliminated and that those commercial fisheries that cannot cooperate be discontinued. The chosen optimums must leave plenty of sea life for the survival of the ocean's biota and its ecological viability. It is clear that such choices will not be made without the facts and force of something like the Sea and Sky Project. These issues seem a bit closer to home when they concern fellow mammals, especially whales.

*What was the value of the blue whale in A.D. 1000?*
*Close to zero. What will be its value in A.D. 3000?*
*Essentially limitless, plus the gratitude of the generation*
*then alive to those who, in their wisdom,*
*saved the whale from extinction.*
Edward O. Wilson

---

# 2

# *Orcas, Right Whales, and Economics*

In a terrifying charge, a bull orca crashed out of the surf and up onto the shore, attempting to seize a young elephant seal only a few feet away. It was unforgettable. I can still see the surge of the onrushing wave he created, its speed, and the surf streaming off his gleaming body. He (I think it was the pod leader, "Mel") looked like a black bus erupting from the water. I had just straightened up from counting the whiskers on a Punta Norte elephant seal (which suddenly seemed silly), about eight feet from the surf, when the attack came. He missed the seal by less than two feet, perhaps distracted by me, and twisted back into the waves.

Whales and dolphins are almost everywhere in Patagonian waters. They include at least six of the eighteen species of beaked whales and huge sperm whales that occasionally strand along the shore and nine of the thirty-three species of dolphins. Among these is the orca, largest of dolphins (called "killer whale" because whales are part of its food), and also baleen whales, including such huge rorquals as the blue and fin whales, and even that virtuoso of cetological song, the humpback whale. Compared to the seals, sea lions, and seabirds, the biomass of these whales is enormous and evidence of their ecological importance. The rare southern right whale (*Eubalena australis*), related to the bowhead and right whales of the distant north, is now the easiest to see and has become the star of a booming tourist industry. In the nineteenth and twentieth centuries, this animal was the favorite prey of whalers. During a prolonged and extraordinarily destructive carnage, North American whalers alone killed sixty thousand, almost all of them.[156] I will introduce Patagonia's orca, a famous predator preying on other marine predators, but focus on the most whaled of whales, the right whale, the larger implications of whaling, and some of the economics of wildlife use.

---

The orca (*Orcinus orca*) has "a reputation of almost mythic proportions," concludes oceans expert Richard Ellis.[157] It deserves that reputation, and none more so than those that hunt along the shores of Patagonia. The orca could deal with a polar bear, a Siberian tiger, or a Nile crocodile like a robin with a worm. It is the largest, smartest wild predator of big vertebrates left on Earth. The fact that it operates in the sea, a frightening environment for humans, anyway, is part of its mystique; but its size and power make it downright fearsome. It will eat almost anything so long as it is big. Fish, squid, seals, penguins, dolphins, whales, even the great blue whale, the largest animal that has ever lived, are all for lunch.

Although they are dolphins, orcas can be over thirty feet long and weigh more than nine tons—twenty Kodiak bears' worth. Their striking black and white coloration, punctuated with a large white spot above the eye, makes them instantly recognizable, although regional populations differ somewhat. They most often live in pods or packs led by an adult male and, at least in the Patagonian populations, including several adult females and subadults; and they share a rich vocabulary of squeaks, whistles, clicks, pulsed sounds, and loud screams. While the relatively large fish-eating pods of the Pacific Northwest are matriarchal, the small groups stalking seals and sea lions along the coast of Peninsula Valdés are led by a single big male, instantly identifiable by a six-foot-tall triangular dorsal fin, and appear to be at least partly territorial.

Think back to Punta Norte, the magnificent beach where Claudio and his students have learned so much of sea lions and elephant seals. From December to early March, about a thousand sea lion pups are born near the Punta. By May, two out of every ten will have been eaten by an orca.[158] Orcas will also have eaten a significant number of young elephant seals and adult sea lions, especially female sea lions. On average, every orca needs to eat from 2.5 to 5.0 percent of its body weight each day. This gives some credence to the theory of several biologists that orcas may be responsible for the disappearance of sea lions in the Falkland/Malvinas. However, orcas are called killer whales with reason. In Antarctic waters, minke whales are thought to make up 85 percent of their diet. Soviet factory ships have reported orca bites on 53.4 percent of the huge fin whales they caught, 29.4 percent of the sei whales, and, amazingly, 65.3 percent of the big sperm whales.[159] The orcas tear off pieces of their prey and swallow them whole. Anyone who has watched piranhas in the Amazon will recognize the style—they don't chew. Sometimes they eat only the tongue of a whale they have killed—like nineteenth-century American buffalo hunters.

Coordinated killing strategies make orcas especially intimidating. For example, they have been seen to surround, then breach, leap from the water, and land on the head of a large whale to block its blowhole, or to hold a smaller one underwater. Jasmine Rossi, an exceptionally accomplished photojournalist who lived in a ranger hut on the shores of Peninsula Valdés for more than a year, watched a female, "Ishtar," and two subadults chase and repeatedly leap on the head of a right whale near the shore of Peninsula Valdés, plainly trying to block its blowhole. In that instance, they let the whale go.[160] The chase appeared to be a training exercise, and many such have been seen; we know of no instances of orcas actually killing right whales near Valdés.

Whalers only occasionally hunt orcas. The Soviets killed 916 in 1980, illegally, and the Norwegians persistently kill them. However, in

1955, the government of Iceland appealed for help in protecting its commercial fishery from orcas. The U.S. Navy responded by destroying hundreds with machine guns, rockets, and depth charges.[161] The orca's reputation as a merciless murderer encouraged the destruction. But then, in the 1960s, orcas came into oceanariums. Skillful trainers were soon persuading them to show off their strength and intelligence in remarkable demonstrations of leaping, diving, and responding to complex cues. Soon people and orcas were rubbing noses, and trainers were riding them in the water. Oceanariums advertised the marvelous animals by name, publicizing their achievements, the births of their babies, and also their tragedies. We began to know them as individuals, and that was the end of the public's perception of orcas as assassins. No navy would think of machine-gunning orcas today, although Mel, the long-time leader of a Peninsula Valdés pod, was immediately recognizable by a bend in his dorsal fin, said to be the result of a rifle shot. Nevertheless, the fact that he and the other members of his wild pod are all known individually reveals that Patagonia's relationship with its orcas is special. A special person led the way.

In 1974, slender, athletic Juan Carlos López, who had trained as an artist and diver, became the first long-term warden at Punta Norte. Fascinated by the visiting orcas, he grew to know each one, named them, and prepared a beautifully illustrated key making them individually recognizable. Over the twelve years that he lived in the warden hut at the Punta with his family, he studied orcas at every opportunity and called their accessibility to the attention of marine scientists around the world.[162] He was especially helped by dolphin experts Bernd and Melany (for whom Mel is named) Würsig, who were then studying dusky and bottlenose dolphins at the nearby WCS whale camp. Gradually, Juan Carlos expanded his educational efforts, speaking to local communities throughout the region, to tourists, and to schoolchildren. He became an eloquent spokesperson for Chubut's marine wildlife, and, thirty years later, as a naturalist and instructor, he continues his educational work.

Punta Norte has been fortunate in having several devoted and distinguished rangers. The latest is Roberto Bubas, who continues Juan

Carlos's interest in orcas. He has attempted to draw them near with music and has shown me a remarkable photo of himself playing a harmonica to a wild orca less than six feet away—not a moment to hit a wrong note. There appears to be no known instance of an orca attacking a human being in nature, even people floundering in the water around sinking boats.

Punta Norte's high tides and steep shingle beaches seem made for orca attacks on sea lions and elephant seals. Although a broad shallow reef, which would seem to keep orcas at bay, fronts much of the shore, several deep channels have eroded through it. When the tide is high, sea lion pups beginning to swim splash about in the shallows. But the channels, especially one of them now called the Orca Channel, make parts of the reef deep enough for orcas to approach the beach, and pup sea lions are easy prey. The area's particular orcas have not only chosen a favorable hunting territory but also developed an unusual technique. They pick sea lions and elephant seals right off the beach at the water's edge—as I learned—sometimes by stranding as much as half of their bodies on the steep shingle.

There are not many orcas in any one place, which is true of almost all top predators. It is a matter of food, that requirement of 2.5–5.0 percent of body weight per day. Nevertheless, concentrations of orcas, thousands strong, have been seen in the Ross Sea, confirming the richness of life in Antarctic waters. No one has ever seen another such aggregation of big terrestrial predators, although small concentrations of big brown bears occur in Kamchatka and Alaska when the salmon are running. On Peninsula Valdés, orcas are now valued as a natural wonder, and their relatively dependable visits and terrifying attacks on sea lions have made them a unique attraction. Getting to know them made the difference. But getting to know the world's most powerful living predator of big animals also makes the proximity of a nursery of slow, apparently vulnerable, southern right whales perplexing.

---

"Right whales" are just right, from a whaler's point of view. They are big, broad, and slow moving. They feed near the surface and float when they die, because of all their fat, which makes "processing" easy. If fat and its oil seem insufficient justification for thousands of human attacks, there were also those bits and pieces used in buggy whips and corset stays to consider. While extensive killing of most species of whales had to await the innovation of the air compressor, which allowed seamen to inflate whale bodies and float them more easily, killing of the right whale did not. Thus, the northern right whales and related bowhead whales were typically attacked first and were almost gone by 1800.[163]

For the southern right whales, those breeding along the shores of Patagonia and southern Africa and feeding in Antarctic waters, most of the slaughter occurred during the nineteenth century, when the major expansion of whaling in the southern oceans began. Recent studies of the history of the southern right whale population suggest that there may have been fewer than eighty thousand, and that, when the main whaling efforts ceased, there may have been only three hundred left (cetologist H. Rosenbaum, personal communication, 2004). Some cetologists believe that a disproportionate number of females had been killed, which might account for the slow recovery. Today there are probably only seventy-five hundred in the world. Of those, about twenty-five hundred visit the shores of Patagonia, mating, courting, and caring for their young, especially in the Golfo San José and Golfo Nuevo. Hundreds of thousands of people come to watch them. Numerically and genetically, the fact that southern right whales are recovering at all is remarkable.

Cetologist Roger Payne, then with WCS, began studying right whales at Peninsula Valdés in 1970 and continues to this day. Individually recognizable patterns of head callosities and raised, roughened patches of skin, which usually appear white because of massive infestations of cyamid crustaceans ("whale lice," which occur on the whales in about the same places as facial hair on human males), made them ideal for study. Many have been photographed each year, usually from an airplane, to note their presence (or absence), the presence of young, association with others, and so on. Vicky Rowntree, who has overseen the work in recent years, says that the 2002 count was made possible by airplanes and pilots provided by the Argentine navy, an encouraging measure of national interest.

As Roger's early studies in Golfo San José progressed, we realized that Peninsula Valdés was uniquely important to right whales. It seemed clear that the Golfo, in particular, should become a reserve. Antonio Torrejon, who headed Chubut's nascent tourism and wildlife department in 1974, readily agreed but pointed out that Chubut's provincial authority over nearby waters was limited. We needed a federal blessing as a foundation for a provincial reserve. I announced I would get it, without the slightest idea how. Luckily, Buenos Aires friends well connected with officialdom helped out. I drafted a memorandum of understanding to be signed by the federal Ministry of the Interior and addressed to the Province of Chubut, supporting and lauding its proposal to make Golfo San José a whale reserve. With advice and help volunteered by Arturo Tarak and other Buenos Aires friends, and a distinct feeling of unease, I flew to Buenos Aires to meet with the Ministry of the Interior.

Astonishingly, we left the Ministry two hours later with a signed memorandum of understanding and delight all around. Arturo and I immediately flew to Patagonia and met with Antonio Torrejon. A few months later, true to his word, he had shepherded protective legislation through Chubut's provincial legislature. In the succeeding years, Roger continued to build a case for protecting the whales, not just in Golfo San José but all the way around Peninsula Valdés, where females were also nursing their babies and adults were courting. With little hope, I subsequently presented the argument to Felipe Lariviere, who had become president of Argentina's national parks. We sat for hours in Felipe's garden and discussed the legislative options and obstacles to creating a reserve in the seas around Valdés—a discouraging prospect.

Years later, we were in for a happy surprise. In a creative move that quickly won government approval, Felipe managed to get right whales declared an Argentine national monument, protected wherever they are in Argentina's waters. Moreover, research scientist José Orensanz, of Centro Nacional Patagónico, eventually succeeded in winning a ban against the use of dredges in Golfo San José, a critical move for both the Golfo's ecology and the whale's peace. Fortunately, no serious threats to the whales have arisen since all that legislation, and there has been little need for its enforcement, but the educational work necessary for its survival is far from complete.

Female southern right whales, which are larger than their male counterparts, can reach fifty-six feet and weigh forty to fifty tons.[164] The twenty-five hundred or so southern rights that now visit the Patagonian coast thus weigh far more than all the elephant seals, penguins, and sea lions together (but less than 10 percent of Argentina's annual fish landings). Unlike the streamlined, fast-swimming rorquals, such as the blue and fin whales, they are very broad, with a head that accounts for almost a third of total body length. It has to be huge to support its giant plankton-straining mouth. Roger says of the humpback whale (*Megaptera novaeangliae*), which is about the same size as a southern right whale, that it, "could have comfortably engulfed a medium-sized car in the cavern of its gape."[165]

Tiny copepod crustaceans and krill are thought to be most of the right whale's diet. When feeding, the whales swim from 4.0 to 5.5 km/h with their huge mouths open, deploying a filtering surface of more than 13.5 square meters of baleen, according to marine mammal biologist Ricardo Bastida. Postulating a concentration of plankton of four thousand grams per meter, he suggests that, at 5.5 km/h, the whale could capture as much as three hundred kilograms of food in an hour. Right whales do little feeding along the shores of northern Patagonia, however. Their enormous size helps them to minimize metabolic loss, and they make up for fasting there by gorging in the subantarctic oceans.

Female right whales become sexually mature at between seven and fifteen years old, males when they are a little older—they have some special growing to do. The testes of male right and bowhead whales are the largest in the animal kingdom, weighing as much as a ton apiece. The reason lies in the nature of their breeding. It is a sperm war. Several males congregate around one or more females and follow them, doggedly for hours, constantly pushing and shoving, jostling for position, trying to mate.[166] Pursued females will often lie on their backs, holding their breath, their bellies out of the water, frustrating the males. However, several males usually do mate with each female, and each male's only chance of producing progeny is to introduce such a great quantity of spermatozoa that his will dominate the race to the female's egg. Chimpanzees have a somewhat similar promiscuous system, which is why male chimps have large testes and gorillas, with one lead male in a harem system, do not.

Despite a surprisingly short gestation of about twelve months (elephants take twenty-two months, rhinos fifteen or sixteen, zebras eleven and a half), the calf gets a good start. At birth, it will weigh about a ton and be thirteen to fifteen feet long. Vicky Rowntree was heartened to find that the population off Peninsula Valdés increased an average of 7 percent each year from 1970 to 1990, and it is continuing to grow. Considering the nearby presence of orcas, that seems remarkable. However, Roger Payne relates a revealing encounter, which suggests that the big, slow right whales are not as defenseless as they appear, at least when in a group:

> I once flew over a pack of killer whales for three hours as they moved along a line of female right whales and their calves. The killer whales approached each mother and calf in turn but I never saw an attack. No wonder. What I did see was each mother flex her body, cocking her tail for a blow toward the closest killer whale.[167]

Unexpectedly, right whales are now being increasingly harassed by kelp gulls in both Golfo San José and Golfo Nuevo. The birds land on a whale's back as it rests on the surface, dig their beaks into its soft skin and blubber, and rip out sizable chunks, eventually creating sores as much as two feet wide. Late one August, as I watched right whales from the shore of Golfo Nuevo (over twenty in sight at once), many with young, seven gulls attacked them repeatedly, causing the great creatures to visibly flinch, dive, and swim away.

Gull attacks on whales are indirectly the result of human actions. They are new, and we can almost date their start. Despite intense observation from 1970 on, no gull attacks were seen until Peter Thomas recorded them in 1984. Vicky Rowntree reported that by

1995, the level of harassment had increased by a factor of five and added that mother-calf pairs were spending 24 percent of their daylight hours "in states of gull-induced disturbance." By 2002, attacks on mother-calf pairs at the WCS station had increased from 12 percent of all five-minute observation intervals to 33 percent.

What changed? Most important, I suspect, is that there are many more gulls, thanks to the abundant year-round food supply provided by fish plant waste at municipal refuse sites. Before the great expansion of local fisheries, and the expansion of garbage produced in growing cities, gull numbers were limited, especially when marine bird and mammal colonies vanished in winter. Not now. And winter is when the whales are near. So disruption of the colonial predator starvation strategy is now also affecting whales. And there are additional considerations and consequences.

Only some, particularly resourceful, gulls attack whales. This is consistent with gull behavior elsewhere, including gull attacks on imperial cormorant chicks at Punta Tombo, terns at Punta León, and zoo ducks. Some gulls are more inventive than others are and may have to be removed. Kelp gull control is also urgent because gulls may be feeding upon infective refuse and might carry new diseases to the whales.[168] Finally, Vicky has noted shifts in the distribution of right whales at Peninsula Valdés and believes that gull harassment could drive the whales from their preferred calving grounds to less suitable areas—also away from the reach of today's economically valuable whale-watching industry.

By the time people began thinking about the economics and ethics of whaling, it was almost too late for right whales. That was not the case with the 130-ton fin whale (*Balaenoptera physalus*), however—although 725,000 were taken during the twentieth century in the Southern Hemisphere alone—or that marvelous giant, the blue whale (*B. musculus*). The blue whale's situation has received particularly penetrating scrutiny and has produced insights into the nature of commercial wild animal exploitation that go far beyond whaling.

In the 1970s, the blue whale faced extinction. This is the world's largest living animal (two hundred tons) and, as far as we know, the

Dee Boersma and Pablo Yorio measure a penguin at Punta Tombo, Argentina.
*Inset:* Magellanic penguin family, Punta Tombo, Argentina.

One of the first pairs of guanays to breed at Punta Tombo, 1969, with their two chicks.

Imperial and rock cormorant colony on an island in Bahia Bustamante, Chubut.

The much maligned double-crested cormorant, New Jersey, U.S.A.

Imperial cormorant feeds chick, Punta León, Chubut.

Rock cormorant chick begs food from parent, Golfo San José, Argentina.

Royal tern and chick, Punta León, Chubut, Argentina.

Graham Harris using "walking blind" at Punta León, Argentina.

Right whales and a southern sea lion, Golfo San José, Argentina.

Albatrosses and giant petrels over the south Atlantic near the Falkland / Malvinas.

Black-browed albatross colony, Steeple Jason Island, Falkland / Malvinas.

Graham Harris and Andrew Taber look down on the world's largest albatross colony, Steeple Jason, Falkland / Malvinas.

Black-browed albatross chick, Steeple Jason Island, Falkland / Malvinas.

Black-browed albatrosses courting, Steeple Jason Island.

Rockhopper penguins returning from feeding at sea, Steeple Jason Island.

Gentoo penguins swimming near Steeple Jason Island, Falkland / Malvinas.

largest since life began. David Attenborough observes, "Its tongue weighs as much as an elephant, and its heart is the size of a car. Some of its blood vessels are so wide that a human could swim down them."[169] In the last century, over 360,000 were killed, and by the early 1970s, only a few hundred were left, like the plight of southern right whales earlier. The Japanese were determined to continue the hunt even to extinction. Then, in 1973, economist Colin Clark of the University of British Columbia decided to calculate, objectively, what action would produce the greatest benefit: stopping the killing until such time as the whale's numbers had recovered sufficiently to permit sustainable harvests indefinitely, or killing the few remaining as quickly as possible and investing the profits in growth stocks. The answer was: Kill them all and invest the profits.[170] That conclusion was so stark, so grossly offensive, that it helped to kill the hunt instead of the whale. It is a classic in the exposition of wildlife exploitation economics.

Princeton's Andrew Dobson has extended the implications of Clark's conclusion: Suppose there were 200,000 blue whales in existence and they were being harvested at a *sustained* annual level of 5 percent. Assuming each whale is worth $7,000, the annual yield from the harvest would be $70 million. Instead, however, suppose we caught all 200,000 whales in one year. We would get a lump sum of $140 billion. If that amount were invested at 10 percent, we would gain $140 million a year—twice that obtained from the sustained use of the resource. Besides, we would have our money up front. "If the growth rate of the exploited population is substantially less than the growth rate of the money earned by its exploitation, it is *always sensible* to overexploit the stock and invest elsewhere [emphasis mine]."[171] And that, alas, is the situation with whales, penguins, parrots, elephants, and virtually all other large, slow-breeding exploitable species. It includes many marketable fishes, especially such long-lived ones as the Patagonian toothfish and New Zealand's orange roughy. However, the investment of the spoils alternative is not "always sensible." Where would the exploiter get 10 percent on his capital in 1930 or in 2002? What market could absorb a large amount of whale meat all at once? And, most important, the "resource" would be lost forever. E. O. Wilson points out that we cannot even begin to calculate what a blue whale might be worth to humanity in the year 3000.[172] Market economics are *never sensible* as a criterion for determining whether or how to preserve nature. And they can be grossly offensive. However, there is now a drive, led by the Japanese, to open whaling again. This time, they say, it will be carefully monitored and controlled.

Commercial takes of wildlife resources, plant as well as animal, almost always lead to overexploitation and near extinction for the exploited species. The resulting income or its prospect generates the power to promote unlimited exploitation for a short time. Few regions illustrate this more clearly than Patagonia, with its overgrazed steppes, fragmented populations of guanacos and rheas, and slowly recovering sea lions and whales. The interdependent threads of the biological fabric that supports such killing, along with its natural variability, usually preclude simple management and, consequently, understandings adequate to manage sustainably the extent of commercial killing that market economies demand.[173]

*In a world older and more complete than ours they move finished and complete, gifted with extensions of the senses we have lost or never attained, living by voices we shall never hear. They are not brethen, they are not underlings; they are other nations, caught with ourselves in the net of life and time, fellow prisoners of the splendour and travail of the earth.*

Henry Beston

---

# 3

# *Report from a Sea and Sky Outpost*

Beneath the clattering helicopter, long-winged black and white albatrosses glided over wind-lashed blue-black water and emerald green kelp as we beat our way west to the edge of the Falkland/Malvinas Archipelago. Isolated islets had toppings of dark brown fur seals, and a whale could be seen spouting in the distance. We were seventy minutes out of the archipelago's Mount Pleasant Airport. Graham, Andrew, and I had flown west over Darwin, the Falkland Sound, and the Hornby Mountains, reveling in views of the islands' strange gray "stone runs," patches of dark diddle-dee bushes, tall waving tussac grass, and white sand beaches. Passing between Saunders and Westpoint Islands, the helicopter swung in the wind over the open ocean. Finally, the spectacular peaks of our destination, remote, uninhabited Steeple Jason Island, appeared ahead.

Within minutes of the pilot's skillful landing in the sharp wind, tents, food, water, sleeping bags, and all the other paraphernalia of foul weather camping were unloaded, and the helicopter departed. Eager to explore, we pitched camp in record time and began hiking. Trying for an impression before the sun set, we walked west, past giant petrel and Antarctic skua colonies, and managed a quick view of the island's famous black-browed albatross rookery, the world's largest. The island was filled with huge birds, all dependent on the bounty of the southwest Atlantic and reflecting its abundance and health. We thought Steeple Jason could be an important and sensitive "well-being station" for the Sea and Sky Project, with sensitive wildlife barometers, and the abundance of the island's seabirds, even at first glance, was good news. We had no idea how quickly that would change.

Steeple Jason is a six-mile-long rocky island dominated by two nearly thousand-foot peaks, the "Steeples," on a steep knobby spine of sedimentary sandstone four hundred million years old. Steeple and its neighbor, Grand Jason, were the first parts of the Falkland/Malvinas ever seen by human beings. They lie at the western end of the 420-island archipelago, 280 miles east of the Straits of Magellan. The archipelago provides a southwest elbow for the Patagonian continental shelf, and it is a biological border marker in the aspirations of the Sea and Sky protection project.

Below its peaks, Steeple Jason consists of rolling heathland and thick shoreside stands of tussac grass (*Poa flabellata*), which can grow as high as eleven feet. Tussac is the biggest native plant in the islands, which have no trees. Southern sea lions rest on Steeple's shores, and whales drift by, straining plankton from the green waters. Fur seals, dolphin gulls, giant petrels, upland and now-rare ruddy-headed geese, Falkland skuas, imperial and rock cormorants are all present, pausing or breeding here. Three species of penguins raise their chicks on the island, and, most important, there is that grandest of albatross colonies.

The Englishman John Davis, captain of the *Desire*, spotted these islands in 1592. Eight years later, on January 21, 1600 (the year *Hamlet* was first performed), they were described in detail by the Dutch navigator Sebald de Weert and were soon called the Sebaldines, but not forever. The British Admiralty sent Captain John McBride on the thirty-two-gun *Jason*, "a fine frigate, coppere sheathed and amply equipped," to do a survey, and he not only sailed through de Weert's islands in 1766 but also "expurged from the British Maps the name of their real discoverer and called them after his ship, the Jason Islands."[174] No humans ever lived here, just truly enormous numbers of trusting seabirds and marine mammals. But, as E. O. Wilson has observed, "Paradise found is paradise lost." Discovery meant slaughter.

For more than one hundred years, Steeple Jason's immense populations of penguins and seals were clubbed, skinned, and boiled. Act II was even bloodier in the Falkland/Malvinas than in Patagonia. When the carnage had converted most of the wildlife into oil and skins, sheep and cattle were dumped on the Jasons, as they had been dumped throughout the archipelago, and the herbivorous assault began. Finally, when the global market for wool and beef declined, making farming on the Jasons uneconomical, and the domestic animals were removed, Grand and Steeple were sold to English aviculturist Leonard Hill, who sought to protect and restore what was left. Upon his death, they were purchased by American philanthropists Michael and Judy Steinhardt. On September 5, 2001, the Steinhardts presented them to the Wildlife Conservation Society as a sanctuary for their original inhabitants, the wild birds and mammals.

In 2000, a once-a-week plane from Santiago, Chile, had brought Graham, Andrew, and me to the Falkland/Malvinas and to introductory meetings in Stanley, capital of the archipelago and its only sizable settlement. It is a prosperous-looking little town of about two thousand of the twenty-six hundred people who dwell on the archipelago. Red-roofed and brightly painted houses, often of corrugated iron, are interspersed with a few masonry government buildings and stores arranged in rows along orderly streets at the edge of Stanley Harbor. A handsome church, Christ Church Cathedral, whose approach is dominated by a whalebone arch, is situated there and surrounded by bright flower gardens in summer. Despite today's serene appearance, the Falkland/Malvinas' economy has always been uncertain. It is, after all, one of the world's most remote inhabited regions, a tough, isolated place to live. Life is dominated by the wind and the sea. Average temperatures range from 36°F (2.2°C) in July to 48°F (8.9°C) in January, and the wind blows persistently, usually at about fifteen knots. For wildlife, both the weather and the isolation seem ideal. Besides seals and sea lions, sixty species of birds breed in the archipelago.

In the early years of settlement, the Falkland/Malvinas did well as a rest-up, refitting, refueling, and repair stop for sealers and whalers and for sailing ships rounding Cape Horn. But beginning in the late eighteenth century and throughout the nineteenth century, elephant seals and fur seals were hunted almost to extinction, and penguins were killed by the millions.[175] Business boomed during the California gold rush, when large numbers of boats from Europe sailed around the horn to reach California. Then, in 1914, when the Panama Canal opened and steam-powered boats came into wide use, refitting collapsed, but farming, especially sheep farming, grew even larger. Livestock had been introduced on the islands almost as soon as they were settled and proved at least as destructive as in Patagonia, but sooner. By 1898, there were over 800,000 sheep on the archipelago, a number now greatly reduced, although sheep raising remains a main industry.

When the effects of bad management and overgrazing began to worry the Falklanders, they responded as farmers do almost everywhere, by blaming the native wildlife—in this instance, the wild upland and ruddy-headed geese (although farmers had introduced hares and rabbits as well as sheep, cows, and horses). In 1905, they got politicians to pass ordinances that allowed the payment of ten shillings per hundred goose beaks. No one knows how many geese were killed, but by 1912, when the bounty stopped, over 500,000 goose beaks had been bought by the government. In 1983, a study found that about 25,000 geese were still being killed annually, and in 1985, another study showed that sheep and cattle ate about 20 percent of the annual production of pasture herbage, while geese took only about 2 percent.[176] Perhaps someday, a farmer will ask himself why, despite large numbers of geese, the pastures were never overgrazed before sheep and cattle were put on them.

The archipelago is self-governing except for foreign affairs and defense. Since the 1982 war with Argentina, it has received much more attention and significant development monies. Nevertheless, and despite all the sheep, 90 percent of the islands' 2003 revenue was based on the sale of fishing licenses—a most precarious foundation. Potential offshore oil reserves may change that mix in the future and pose new dangers to wildlife. There is little official conservation work. Most such work is done by a small but outstanding nongovernmental conservation organization, Falklands Conservation (FC), which had conducted a critical census of many of the major wildlife populations on the islands before our arrival in Stanley. To my relief, Falklands Conservation welcomed Graham, Andrew, and me with as much courtesy and goodwill as Argentines had been welcoming their staff. Its excellent wildlife studies made it clear that FC would be an important collaborator, not only in archipelago programs but also in the overall Sea and Sky plan.

The Jason Islands gave us fresh insights on Southern Cone conservation, specifically on our thinking about a Sea and Sky reserve in the southwest Atlantic. They also gave us a chair at the south Atlantic planning table, a direct connection with the British scientists working in the region. We began thinking about the feasibility of using Steeple and Grand Jason as part of a network of observatories to monitor the fortunes of coastal and oceanic animals of the south Atlantic. The islands' albatross and penguin colonies seemed likely to prove particularly important to understanding what was happening between south Atlantic fisheries and wildlife. After two days in Stanley, we arranged for a helicopter to take us to Steeple Jason.

It was cool and wet when we crawled out of our tents on the morning of our first full day at Steeple Jason. We were grateful and a little surprised that the tents had withstood the pounding of the night's winds. We had joined up with wildlife cinematographers Mark and Jane Smith, who were camping on Steeple to finish a documentary on its clever striated caracara (*Phalcoboenus australis*), perhaps the largest population of the rare "Johnny rook" remaining. Behaving like monkeys in hawk suits, these fearless birds take any opportunity to rifle a backpack, steal a sandwich, or attempt to disassemble a helicopter.

Steeple Jason's stark beauty, unpeopled vistas, and surprising size surpassed our every expectation. After inhaling coffee and adjusting boots and jackets that first morning, we split destinations. I hiked toward the celebrated albatross colony. The damp landscape sparkled. Large birds were everywhere—giant petrels, caracaras, upland geese, skuas, gulls, and terns. Petrels and prions, the petrels' little blue-gray relatives, were nesting in burrows on the sides of the "steeples." The albatross colony was about a mile from camp and screened by a dense tussac forest. I alternated pushing through the all but impenetrable grasses with jumping from rock to rock along the boulder-strewn shore. Magellanic oystercatchers and kelp geese, the pure white male contrasting with his gray-brown striped mate and gray goslings, fed on water plants in the tidepools. Johnny rooks followed my progress with seemingly ill-disguised hope that I would fall off a rock and break my neck.

Abruptly, the tussac gave way, and I found myself on a sweeping plantless plateau crowded with thousands of albatrosses. The swan-sized parents tending chicks sat on foot-high nest mounds, while onlookers and those on rest breaks socialized on nearby rock outcroppings. All appeared very dignified. I could see more than a mile of the giant birds—an overwhelming spectacle. In fact, the colony stretches a staggering two or more miles along Steeple's southwest coast.

The air was saturated with albatrosses. It was mesmerizing—thousands of huge birds soaring and gliding, over, under, and between each other, with rarely a wing beat. Each bird's eight-foot wingspan was fully extended, except immediately before landings, which were not always graceful. At high speed, their knife-like wings created an unnerving whistle, like a sailboat's stays in the wind. One brushed across my hat, leaving a mixed sense of narrow escape and accepting affinity. After several counts, I estimated that there were approximately fifteen thousand in the air, which meant nearly twenty-three miles of albatross wings over my head at one time.

The beauty of the black-browed albatross's soaring has moved many to descriptive flights. In *Oceanic Birds of South America*, Robert Cushman Murphy wrote:

> Every movement revealed the constant, delicate reactions of the mechanism of balance—the gentle, almost unnoticeable rocking and seesawing of the wings with the bird's body as the fulcrum, the gauging of the angle of the wing-axis with the horizon according to the sharpness of the turn, and the feat of "shortening sail" at a critical moment, the last capability being due to the structural advantages of jointed planes, which man has thus far been unable to build into his imitation aircraft.[177]

In *Eye of the Albatross,* Carl Safina, a musician in a former life, writes:

> An albatross is a great symphony of flesh, perception, bone and feathers, composed of long movements and set to ever-changing rhythms of light, wind and water. The almost overwhelming musicality of an albatross in air derives not just from the bird itself but from the contrapuntal suite of action and inaction from which this creature composes flight. It drifts in the atmosphere at high speed, but itself remains immobile—an immense bird holding stock-still yet shooting through the wind.[178]

Albatross wings are relatively flat with little camber. The feathers of their wing tips lack slots like those of birds with broad wings, such as pheasants, broad-winged hawks, and condors, which help in takeoffs and landings and provide for quick turns. But narrow, flat albatross wings reduce drag. An albatross can glide at high speed but is not very maneuverable. However, its ability to take advantage of strong winds is the secret of albatross life. The birds live on their wings. They are capable of slicing through a sixty-mile-an-hour gale and gliding a hundred miles without a wing beat. Theirs is the ultimate gliding design, the shapes and skills descended from evolutionary urgencies of old winds, the tempests of millions of years past.

Close up, the black-browed albatross (*Thalassarche melanophris*) is black and white, mostly white. It has a broad, Henry Fonda–like forehead, large dark eyes, and from just in front of those beautiful eyes and extending a short way behind them, an eye-shadow-like shading of black. Its bill is big, hooked, and a delicate orange-yellow. Standing high on their nests and clapping their beaks at any perceived danger, the downy half-grown chicks are the living embodiment of Big Bird, except that they are gray. There is something about their combination of form, fluff, and behavior that makes humans grin and feel protective toward them. They seem so comically ready to defend themselves but so vulnerable and childlike. If only there were some way for more human beings to experience such scenes, without destroying them.

Each pair's nest, a rock-hard stack of dried mud, dung, and debris, is slightly beyond the reach of its neighbor's beak, which imposes a geometric order on the colony. Nic Huin of Falklands Conservation had just counted 157,000 breeding pairs on Steeple before I got there, and there were numbers of nonbreeding "visitors." There were also 100,000 or more chicks, so the whole elegant spectacle exceeded 400,000 albatrosses. They ignored me. I was not a "clear and present danger," as I had been to Punta León's cautious gulls and terns, just an inconsequential spook in the midst of a colossal avian convocation that had never known mammalian predators.

In contrast to the magisterial appearance of the sailing albatrosses, a robin-sized, dull brown tussac bird (*Cinclodes antarcticus*) searched the colony's debris and hopped around my boots. Eventually, he pulled my shoelaces and perched on my tripod to preen his short wings and defecate on my sleeve. Such fearless behavior was typical of almost all Falkland/Malvinas wildlife when the islands were discovered. Geese were so tame that sailors would simply walk out with a handful of rocks to kill some for dinner. The one significant land mammal, a small fox related to Patagonia's gray fox and called the warrah (*Dusicyon antarcticus*) was so trusting that all were quickly killed. As Graham sat observing a nearby colony of gentoo penguins (*Pygoscelis papua*), he was shocked and delighted when chicks clambered into his lap. How easy it must have been for the first people in contact with such naive animals to kill them.

Small parades of self-absorbed knee-high rockhopper penguins (*Eudyptes chrysocome*) marched through the colony single file, wending their way between the big albatrosses sitting on their nest mounds above them as though on thrones. The rockhoppers nest in separate colonies but also among the albatrosses. Nic Huin had recently counted 89,000 in this mixed colony. In the 1930s, the total rockhopper numbers in the Falkland/Malvinas were about 1.5 million breeding pairs, but they had declined to 263,000 pairs by the mid-1990s and then reached 272,000 breeding pairs in 2002.[179]

In the 1860s on Steeple Jason Island, both rockhopper and gentoo penguins had immense colonies before they were killed and boiled for their oil. Oilers calculated that 8 penguins would produce one gallon of oil. From 1864 through 1866 alone, sixty-three thousand gallons of penguin oil were transshipped through Port Stanley (the result of boiling at least 504,000 penguins). It is estimated that about 2,000,000 gentoo and rockhopper penguins were killed for that industry during its sixteen-year life, Act II at its height.[180] When Graham, Andrew, and I made that first visit to Steeple Jason in January 2000, old try-pots could still be found rusting on the island's shores among countless penguin bones.

All around me, albatrosses were bowing, billing, and posturing, while yellow-crested rockhoppers gathered in groups among them, shouting at the top of their lungs, squeakily trumpeting their hilarious bad-hair day, their well-being, their territorial rights, and their oneness with their mates and chicks. Imperial cormorants, serpentine necks swaying from side to side, bright blue eye rings and golden caruncles at hormonal peak, formed yet another strata in the huge colony's society, voicing low grunting growls and, from time to time, regurgitating fishes to their downy black chicks. Johnny rooks, those strangely clever hawks, and Antarctic skuas (*Catharacta antarctica*), which are hawk-like gull relatives, constantly patrol the rookeries, seeking opportunities to eat unprotected eggs and chicks.

Once adult, albatrosses can live more than fifty years. A New Zealand royal albatross is the record holder. "Grandma" did not fail to return home to her colony until she was sixty-two, in 2001. However, the vast majority of immature albatrosses, like the young of virtually every wild species, do not survive to breed even once. They do not begin to breed until they are about eight years old. So longevity and memory are big factors in albatross success. If you were the world's greatest animal traveler, as they are, it would help to be able to keep track of where you are and to remember where you have been.

Thirty-two Falkland black-browed albatrosses have been followed on their food-seeking flights with global positioning and activity sensors. Trips by birds with nests averaged 6.8 days, nonbreeders 127.5 days. Their foraging areas ranged from 15,205 to 19,578 square miles by incubating birds and up to 231,362 square miles by nonbreeders. By the time an adult like Grandma dies, it has traveled several million miles. These are not point-to-point flights like those of many migrating birds. Albatrosses are circling, searching for food, and exploring thousands of miles of open ocean with no landmarks obvious to us. Yet when they wish to get back to their chicks, they know the route straight home. One tagged wandering albatross feeding a chick in a nest flew 9,000 miles between feedings.

How do all these birds compare with people when it comes to food biomass? Safina reports that "the seabirds of the Northwestern Hawaiian Islands require an estimated eight hundred million pounds [400,000 tons] of prey annually—largely fishes, crustaceans, and squids—perhaps two-fifths of the annual production of those animals in the ocean region. Of the amount consumed, albatrosses probably take half."[181] The total seabird "take," then, is about .0047 of the reported commercial fisheries catch and largely of different species.

---

The seabird slaughter of the 1800s up to the early twentieth century was astonishing—fully the equal of whaling but much less known. Eggers collected so many albatross eggs on Laysan, in the Hawaiian Islands, that they used ore carts to transport them. The eggs were wanted for the photographic industry, which used them to make the albumen prints that were popular at the time.[182] Mind you, these are birds that lay but one egg a year and don't breed at all until they are eight years old or older—a breeding age as old as a whale's or rhino's.

Even more devastating was the feather business. The Japanese killed millions of albatrosses for their feathers: "In 1902 a visitor to Midway Atoll found thousands of seabird bodies killed by the Japanese. He wrote, 'Everywhere on Eastern Island great heaps, waist high, of dead albatrosses were found.'"[183] After the Japanese feather hunters were finally evicted from Midway in 1903, they landed seventy-seven men on Lisianski Island (between Midway and Laysan Islands in the Hawaii archipelago) in 1904, and killed 284,000 birds in six months. Even after President Theodore Roosevelt, prompted by the Audubon Society, declared Laysan a sanctuary for seabirds in 1909, Japanese poachers landed on the island and killed over 128,000 birds. Eventually, only an ongoing American presence stopped the killing.[184]

All twenty-one species of albatrosses, versus about five in 1994, are now listed as threatened. Fishing operations are killing the birds by the thousands, as bycatch. In the Southern Ocean, an estimated 44,000 albatrosses were killed each year in the late 1980s. A more recent compilation put the toll as high as 330,000 seabirds, including 46,500 albatrosses, 7,200 giant petrels, and 138,000 white-chinned petrels killed between 1997 and 2000 on "pirate long-lines" in the Southern Ocean.[185] A 2000 census revealed that the Falklands black-browed albatross population had lost about two birds an hour every hour for the previous five years and that the rate was increasing. The culprits include the long-line fisheries, especially those for toothfish, from Spain, Korea, Japan, and Argentina, and, just as seriously, demersal trawlers.

In 2001, the Australian minister of the environment called a major conference of fishing nations, and, on June 19 of that year, a treaty to protect albatrosses was signed in Australia by all the nations surrounding the southern seas, but without providing for independent observers. There is an unwillingness, even in Australia, to close ports to Japanese long-liners, which are said to be the worst offenders. About $40 million in harbor and docking fees is involved. In 2004, the Agreement on the Conservation of Albatrosses and Petrels (ACAP), which had been signed by Australia, New Zealand, Ecuador, Spain, and South Africa, came into force. It requires its signatories to protect the birds' nesting grounds and to take measures to reduce seabird bycatch—but the main culprit nations have not yet signed.

Although the Sea and Sky Project has gained momentum, it is still in the political negotiation and fact-gathering stages, a long way from implementation. The effort is focusing on three objectives: creating open-ocean protected areas, especially corridors for migrating penguins, squid, albatrosses, and elephant seals; exploring enforcement alternatives; and developing an overall communication strategy. Scientific research, zoning, and developing a legal framework for implementing the plan are but a few of the tasks under way. The effort is spurred by devastating news.

The suspicion that Steeple Jason might be a particularly sensitive barometer of the health of the southwest Atlantic ecosystem has borne bitter fruit. Since our visit in 2000, its great black-browed albatross colony had lost 88,150 albatrosses (28 percent) by 2004. Graham, Andrew, and I fear that we may have seen the birds during the last period when their colony was healthy. As a result of paralytic shellfish poisoning from harmful algae blooms, those little self-absorbed rockhopper penguins have declined by 66 percent.[186] We cannot yet assign cause to the increase in such blooms. It may be tied to the major ecological disruptions caused by overfishing and, possibly, by the dumping of bycatch and the fishery operations themselves. Steeple Jason's seabirds, though, are telling us that the fishing industry is creating silent skies and empty seas.

*Our optimism stems from the realization that greatly increasing the areas protecting biodiversity represents a clear and achievable goal . . .*

Stuart Pimm, 2001

# The Scene Ahead

The wonder of the Southern Cone is that, with the exception of the Falklands fox, the entire cast of creatures left to us by the Tehuelches, before the nineteenth- and twentieth-century slaughters and settlements by Europeans, is still onstage, every single species. Yes, they are greatly diminished, but you can still see them all, except for the Tehuelches themselves. The shape of the scenes to come is now to no small degree in our hands. What are the prospects for a viable future?

Globally, the impacts of increased human population and development are expected to result in massive species extinctions on land and failing ecosystems in the sea. In Patagonia, effects of some ugly scenes from Act II drag on, and new threats to wildlife are arising. The *2004 IUCN Red List of Threatened Species* reports that 20 percent of the world's mammals, 12 percent of the birds, and between 3 and 31 percent of the fishes, reptiles, and amphibians are threatened with extinction.[1] Nevertheless, wildlife in the Southern Cone is hardy, human population is sparse, and the people's commitment to the environment is growing. Besides, Patagonian penguins, sea lions, and whales are now starring in science, television, and news stories throughout the world. Although some are on the *Red List*, their destruction would no longer be unnoticed, as it was a few decades ago. Act III has new directions.

Within the past two decades, scientific studies of Southern Cone wildlife have stimulated hundreds of articles and helped to attract thousands of tourists. Virtually all Patagonian shops now feature wildlife motifs and artwork, while hotels and travel agencies promote wildlife tours. Anti-whaling sentiment has become at least as strong in Puerto Madryn as it is in Boston. The Patagonian Coastal Zone Management Plan and its offshoot, the Sea and Sky Project, are big and imaginative, the first under way and the second in serious negotiation. The new triumvirate of science, public participation, and tourism aims at saving not only the wildlife we see but also the sources of its support that we don't see.

And about that rancher's advice in the form of an observation, that I could put "all the conservationists in Patagonia in [my] old Ford and still have room for a large dog in the back seat"—not nowadays. Today, a Chihuahua couldn't find a seat in a very big bus of Patagonian conservationists. There is a however, however.

Southern Cone wildlife faces a rising level of incompatible economic activity and trade policies that undermine environmental protection. The incubus is not just increased local exploitation but the Tragedy of the Commons reconstituted as the Detachment of Globalization. Investors living far away in Rome, Madrid, and Chicago affect the prices of wool, hence sheep numbers on the fragile pastures of the steppes. Foreigners also affect demands for oil, minerals, tourism, even water and roads, hence new threats to wildlife in the altiplano as well as on the steppes. They control marketplace conditions for fish and squid, and therefore the prospects of the wildlife of both coast and sea. Alien investors may even determine where hotels are built, initiate mariculture, and begin new industries—often without serious public review.

These threats are not new and not peculiar to Patagonia. International trade and investment have a long and often constructive history; but now they are much bigger and more widespread, and they are affecting a wildlife whose populations have become fragmented, constricted, and far smaller. Whatever the economic virtues of market-based globalization, the detachment of financial control from environmental responsibility and caring will usually be a disaster for wildlife—and even that strong statement is quibbling. We could just as easily describe a Catastrophe of Commercialization, a Curse of Short-Term Profit, and a Disregard of Consequences, as a Detachment of Globalization in the Southern Cone. Worse, in an era of no-fault economics, to be antiglobalization or antidevelopment is to be on the wrong side of history and of Argentine perceptions. So what is needed is more development—of protected areas and zones, conservation organizations, environmental education, and informed control, oversight, transparency, and accountability—and globalization of conservation support. Today, all nations share and affect Earth's wildlife.

Although saving wildlife is most effective at the landscape level, all local experts of my acquaintance believe that, in view of current threats, most steppe wildlife protection will have to take place in sanctuaries largely separated from human developments. The current cultural climate needs time to evolve, time for people to accept that saving their environment is saving their jobs, that habitat destruction is a human problem, not an environmental one. Without separation of wildlife from many kinds of development, too much habitat and wildlife will be lost too soon for changing public views to take effect. Only a little more than 1 percent of Argentina is given over to protected areas, and my Tanzanian parks colleague would observe that, in Patagonia, much of what is protected is ice and rocks. If coastal mammal and bird colonies are to withstand increasing economic activity and competition for food, they must have not only more protected areas on land (protected provincial reserves probably total less than 0.003 of the Patagonian coast), but safekeeping of their critical ocean feeding grounds, as well. Up on the steppes and Andean altiplano, sanctuaries focused on the lagunas are the most practical recourse for threatened wildlife, but on the shore and in the sea, *zonal* systems of protection that restrain overfishing and choose wise shore development are the best option.

Most of this book has been about animals, their lives, risks, and how they deal with each other and we with them. This is fundamental information, mostly missing until recently, and all too easily ignored in humanity's thickets of generalization and self-concern, even when we think about how to do conservation. Because there is no good way to ignore those brambles, I've tried to boil them down to basics. For example: Why, some ask, can't wildlife be protected more generally through what is called sustainable development? Why can't an equitable balance between human and wildlife needs be reached outside parks and reserves? And, even if sanctuaries are a first principle, who should apply it?

Designating protected areas, providing for their management, and enacting wildlife laws are not made easier by several well-meant but vague conservation ideas that almost universally influence government decision making. Chief among those cloudy concepts is that of "sustainable development," which is sometimes used to argue *against* protected areas.

The phrase *sustainable development* was popularized by the World Commission on Environment and Development in 1987. The original definition was "development that seeks to meet the needs and aspirations of the present without compromising the ability to meet those of the future." It seems compelling. However, insofar as wildlife is concerned, it is also ambiguous, for it usually ignores the preservation of nature and ecosystem sustainability. Worse yet, its proponents rarely concern themselves with a consideration of carrying capacity (the maximum population of a species allowed by an environment) or the fact that the vast majority of wild species and their broader ecological services are neither harvested nor credited in human economies. The trouble is that "sustainable development" is interpreted by many as meaning development that will create more and more income for more and more people. (We are going to catch more and bigger fish every year.) Disingenuously, few who use the term in relation to wildlife conservation define what they mean.

Outside of tourism, examples of sustainable development that are consistent with wildlife protection are uncommon. It is more realistic to work for sustainable *use*. Using penguins for tourism rather than for gloves and pig feed requires no killing of penguins and, to a point, no increase in carrying capacity. That point is reached when tourist disturbance exceeds the tolerance of penguins to withstand it, or, perhaps, when we compare the public support costs of tourism and the lost opportunities of those who wish to make penguin gloves or overfish penguin food.

Although tourism's benefits tend to be distributed throughout their local communities, in contrast to those of oil or commercial fisheries, tourism is not nearly so profitable, and its future is wholly dependent on government nature protection—and control of the fisheries. When the investigative phase of the Patagonian Coastal Zone Management Plan was completed in 1996, three figures stood out: The oil industry was bringing in about $1 billion a year, commercial fisheries about $500 million, and tourism about $90 million. All those figures are much higher now, but the relative proportions are much the same.

Nevertheless, wildlife tourism is more easily sustainable than most wildlife harvests, such as shooting guanacos for skins and meat. Such harvests are capped because no natural wildlife population can support persistent growth in its "harvest," as we saw in the discussion of whaling. However, there could be substantial sustainable use with the development of extractive reserves, as suggested in Chapter 8, "A Once and Future Patagonian Steppe." Besides, the taking of calculated numbers of invasive species and predators, such as pumas, boars, hares, and foxes, might provide benefits for beleaguered native herbivores such as maras and choiques.

Except for tourism and scientifically managed harvests of certain species, economic development is best conducted away from protected wildlife populations, but keeping such development away from supportive ecosystems is not easy. Although Puerto Madryn's big aluminum plant has little direct effect on wildlife, its indirect effect, in the form of additional people, demand for more living space, and related developments, is enormous. The plant's positive human impacts in terms of jobs and community space are also large. But such impacts will overwhelm wildlife populations and habitats except where specific protections are established. Sanctuaries are basic. Besides, ongoing development of any kind can run afoul of fluctuating market forces; a simple change in fashions or hobbies can mean a complicated death for chulengos or parrots.

Yet another slippery shibboleth hiding in the ignorance thicket is the attractive aim of "achieving a balance with nature." As the book *Life of Pi* suggests, it is hard to achieve equity in a small boat with a tiger.[2] How will the needs of sheep and guanacos or fishing boats, pumas, albatrosses, and penguins be balanced? I suppose that "achieving a balance" could mean agreeing upon the population and habitat sizes we will allow animals that compete with us, providing those allotments meet the animals' biological necessities. (The world's seabirds are said to eat about seventy million tons of seafood annually, about 0.4 percent, in competition with human fisheries. Harp seals take about four million tons in the northwest Atlantic each year, and, of course, there are those whales and elephant seals and hundreds of other creatures with which we share the global dining table.)

Figuring out how to meet the needs in reserves of three or four landscape or flagship species, such as guanacos, rheas, and flamingos on the steppe or altiplano, as stand-ins for their larger communities, can give us guidance. In this way, the balance concept does become a tool for conservation. But scales have pans on either side. What numbers and resources do we permit ourselves? E. O. Wilson gently suggests 50 percent, "half for humanity, half for the rest of life."[3] Wilson's suggestion is questioning rather than calculated but a useful point of departure, and almost certainly too small for "the rest of life" if we expect it to sustain our biosphere. In any event, nature is finite, while human aspirations are not, and that point reemphasizes the need for stoutly protected wildlife zones and sanctuaries—and restraining our own numbers.

That we must share some portion of the earth's resources with animal life, as it seeks to survive in the crevices between our ambitions, is reality. If flamingos and the wonderful wildlife community associated with them are to exist in the altiplano, we will have to secure their specialized lakes from exploitation. If guanacos, tortoises, and burrowing parrots are to persist, we will have to share steppe plants, some of the agricultural crops that have replaced their original foods, and sandstone cliffs. If penguins, seals, and sea lions are to journey into the future with us, we must ensure their use of shorelines, marine invertebrates, and fishes—and accept lost private opportunity as a premium. And what about pumas and culpeos? Someone must make decisions, hundreds of them. So who decides what zones, sanctuaries, and reserves will be created?

"Parks," "reserves," "refuges," and "sanctuaries" are part of a lingua franca of nature care, but to be created and maintained, the entities themselves must be embraced as morally good. There must be a marriage between values, vision, and policy. Despoiling a sanctuary by grazing goats and cattle in it should be perceived with the same outrage as stealing land. Most Patagonians (and many other people) are a long way from that view.

While everyone can agree that reserve designation and wildlife and habitat preservation require special expertise and long-term commitment, such qualifications are not easily met by even the most interested elected officials when they have little appropriate training and are put out of office every few years. Those qualifications are even less likely to be found in the changeable private hands of the 5 percent of the populace who own most of the vast steppes, to say nothing of the complex altiplano or the coast, and scientists will rarely have a chance to make reserve designation decisions independently. Nevertheless, an especially popular maxim, affecting both conservation action and the job of creating and caring for parks, is that "local stakeholders must be a part of any attempt to preserve nature." Where does that generality lead?

A 1993 conference of the IUCN (World Conservation Union, formerly International Union for Conservation of Nature and Natural Resources) concluded that "the key to protecting a cherished landscape lies within the communities that call it home."[4] Those communities, of course, are usually the people who live in the area or nearby and whose practices and opportunities may be affected by protected areas or legislation. The notion is that such communities are dependent on the "healthy survival" of such landscapes and, consequently, that they are the best qualified and motivated people to manage and protect them. This is good politics, wonderful sociology (in some regions), and one of the conservation establishment's most confused non sequiturs. Do we really think conservation is that simple?

Wildlife reserves must not be left in the hands of individuals whose only credentials are that they happen to have moved in next door, often recently. In the Southern Cone, such people are commonly the major sources of habitat destruction, invasive species, pollution, and overhunting. Long-tenured indigenous people are mostly gone, and those who care about preserving cherished landscapes are rare. When they do work to care for their land and resources, they tend to be powerless in the face of globalization and broader economic forces. What stakes in the proceedings do the wildlife and those who care about it have? Or the regional and national citizenry beyond the locals? From the standpoint of the viability of almost any environment or commercially harvested wild "stock," the requirement that its designated stakeholders be committed to its long-term survival is fundamental. What is essential is to bring together diverse and appropriate stakeholders in a science-based process of solid information, consensus building, and analytic deliberation—and to restrain those bent on ignoring public values for short-term personal gain. In Patagonia, this means that the views of the larger populace, not simply the scarce inhabitants of particular areas, must be taken into consideration. Along with the vast majority of the world's wild creatures, most of the Southern Cone's wildlife can be preserved only for a public good, not a private one—for ethical, aesthetic, and ecological reasons.

Is a Taiwanese fishing company a proper stakeholder in the Mar Argentino, an Austrian sweater maker a stakeholder in the care of Patagonia's steppelands, a Chilean mining company in the exportation of Bolivian altiplano water? Recently, an association of snowmobile manufacturers, asserting stakeholder status, helped override a U.S. National Parks decision prohibiting snowmobiles in Yellowstone and Grand Teton National Parks. Where the survival of wildlife is concerned, one should no more leave the future of protected areas to the inexpert, or to self-seeking commercial interests, than to do that

still not regarded as "proper science." In the United States, if thought of at all, conservation biology was considered an "applied" discipline, not of the same academic stature as theoretical biology, hence not for ambitious academicians. That deformity of priorities, and distance from conservation action, which ecologist Dan Janzen has likened to "studying the nature of fire amidst a forest fire," is far from cured. Even now, few leading positions are available in academia to scientists actively participating in conservation programs, as compared to those who focus on theoretical science. Understandably, many top-flight Argentine biologists wanted to win their spurs in accepted university fields rather than fighting for conservation with politicians and developers in low-paid jobs in the hustings. Some feel that this bias against conservation science has contributed to a lack of strong Argentine national park policies; it has infected many other Southern Cone reserve systems as well.

In Bolivia's altiplano, the relationship of the growing tourism industry to the peerless Laguna Colorada is largely determined by 4x4 tour drivers quite beyond the control of the ranger staff. The great majority of the country's magnificent Andean lagunas are totally unprotected, and its parks management has rarely been strong. In contrast, the Argentine national park system was once among the finest in the world; its 1981 park book, *Los Parques Nacionales de la Argentina y Otras de Sus Áreas Naturales*,[7] is one of the best such publications I know. Nowadays, the national parks administration is simply a division within the tourism ministry, not a top-priority agency for irreplaceable national treasures.

Despite the problems and deficiencies, it seems necessary for most major sanctuaries in the Southern Cone, ultimately, to be run by national or provincial governments. They are a public trust, and the government probably has more endurance than private or community-run alternatives. Moreover, there are proven ways of strengthening the system. If governmental reserves choose to collaborate with the growing community of nongovernmental conservation organizations as a matter of policy, solutions for many sanctuary shortcomings will be found. The alternative of privately owned reserves can rarely promise continuity, unfortunately—to say nothing of protecting viable populations of wide-ranging wild species. Although the supervision of profitable fisheries can be supported by user fees, global experience suggests that most sanctuaries require ongoing public help. To paraphrase: Government parks and reserves are the worst of all forms of conservation except for all the other forms that have been tried from time to time—and the costs are reasonable. E. O. Wilson has observed, "The protection and management of the world's existing natural reserves could be financed by a one-cent-per-cup tax on coffee."[8]

The "boundless" Patagonian wilderness that Darwin and Musters knew is nearly gone. It has been converted to boundless emptiness and sheep. Yet restoring its beauty and mystique is one of the most intriguing conservation challenges anywhere. Andres Novaro and Susan Walker call it "rewilding Patagonia," and it can be done. My plea for future populations of 2,000,000 or more guanacos, 4,000,000 rheas, 500,000 sea lions, and 3,000,000 penguins is not an idle exhortation. They are feasible numbers and provide measurable objectives—but only within a context of public support and management. New partnerships between government reserve agencies, provincial and federal, and nongovernmental conservation organizations, such as the Patagonian Coastal Zone Management Plan, are a promising model. More experiments are on the way in the Sea and Sky Project, in efforts on the steppes by Patagonian provincial governments, and in other projects by Fundación Patagonia Natural and Fundación Vida Silvestre Argentina.

A recent development in the Southern Cone is the private acquisition of large, important properties for conservation by conservation organizations and individuals. Until recently, such groups as The Nature Conservancy, Conservation International, and WCS had confined their acquisitions of land for reserves mainly to north temperate and tropical regions. With the purchase of extensive properties in Chile, such as Pumalin and Chacobuco, and of Monte León in Argentina, by Doug and Kris Tompkins; Steeple and Grand Jason Islands in the Falkland/Malvinas by Michael and Judy Steinhardt; and the vast Trillium properties in Chilean Tierra del Fuego contributed to WCS by Goldman-Sachs, the potential of such generosity to help save Southern Cone wildlife has changed. The best formulas to ensure public involvement in these efforts, however, are still developing.

There are promising smaller efforts, too, such as an organization of Rio Negro province residents trying to preserve the great barranquero colony at El Cóndor and the fascinating reintroduction of the Andean condor in the Somuncará. Concerned groups of people promoting the conservation of their region's wildlife is sustainable development in the best sense. The Concern of the Citizen is a powerful antidote to the Detachment of Globalization. In Aldo Leopold's vision, wilderness preservation is not denial of "progress" but a new and advanced fulfillment of it.[9] In Patagonia, a corner is being turned.

Volunteer conservation organizations are now inspiring nature protection over much of the world and partnering with government officials to provide stewardship. In the Falkland/Malvinas Archipelago, the small but vibrant nongovernmental organization Falklands Conservation plays the leading conservation role. Argentina has now developed both a cadre of cooperative NGOs and the interest of international organizations with local expertise. Charitable conservation organizations offer a way of putting society's interest above that of the individual entrepreneur or a thoughtless local community. They give wildlife a voice. Fortunately, many of the dilemmas that make wildlife preservation so discouraging in overpopulated, less-developed tropical regions are not yet an issue in the Southern Cone, except for that of the worldwide deficit in environmental education.

Overlooking the broad Golfo Nuevo, on a high bluff at the south edge of Puerto Madryn, stands one of the most dramatic efforts to teach people about their environment anywhere. Called EcoCentro, the new museumlike structure is a beautifully composed combination of art, science, and conservation devoted to the marine life of the Patagonian coast. Its exhibits are handsomely designed expositions, videos, and models featuring the lives of orcas and elephant seals, penguins, fish, sea lions, and other regional oceanic creatures, and the life of the ocean itself. Theaters present inspiring films of marine life, alternating with musical performances and art exhibitions. The building has large windows, a tower, and a great deck all looking out over the Golfo, where right whales are commonly seen. It is an imaginative and theatrical effort to inject environmental consciousness and appreciation into Patagonian culture. It was created through the efforts of Alfredo Lichter, the same Buenos Aires scientist whose book on Southern Cone marine mammals caused me to write about coastal conservation and led, indirectly, to the Patagonian Coastal Zone Management Plan. It is sorely needed.

In the theater of Patagonia's human-wildlife relationships, environmental ignorance and unintended consequences play an ever larger role, even in Act III. No wonder—there are more people. By discharging an endless stream of garbage, we feed aggressive kelp gulls, encouraging them to multiply and expand their depredations on terns, cormorants, penguins, and even right whales. By introducing hares and rabbits, we further degrade the pastures of the steppes and subsidize outsize populations of foxes, pumas, and buzzard eagles, with the result that there are more predators to eat both our sheep and the diminishing populations of Darwin's rheas, maras, vizcachas, tinamous, waterfowl, and guanacos, and even to break down colonial bird defenses. Our introduced European boars and domestic livestock degrade "protected" reserves, and the boars eat both flamingo chicks and eggs and even the seeds of the giant, redwood-like alerces.

Every Southern Cone conservation scientist has his or her "worst fears" about the future of the region, about the international deficit in environmental education. "What would happen," Dee Boersma worries, "if a major anchovy fishery began?" The history of the Humboldt Current fishery off Peru suggests that the coastal penguin, cormorant, and tern communities and much of the fishery for larger fishes might collapse. What would happen if maricultural developments for salmon and other predatory fish were approved? Experience suggests that they would deplete native fishes to feed farmed fish; seriously affect native seabirds, marine mammals, and fisheries; and become a major source of pollution and, perhaps, fish disease. "Will the wild boar extend its range south from Rio Negro and affect coastal bird colonies?" Graham Harris asks. "How will increasing numbers of pumas be managed, from the standpoint not only of sheep but also of today's reduced numbers of guanacos, maras, and rheas?" frets Ricardo Baldi. "Will a lack of understanding between Argentina and Great Britain condemn the southwest Atlantic's unique world of sea and sky to ecological destruction?" worries Claudio Campagna.

Although many of Act III's issues are old ones, the drama's direction has moved from one of wildlife loss and destruction to one of appreciation and preservation. Its animal protagonists survive; its heroes are newly thoughtful public officials, local citizens, scientists, and deeply committed conservationists. From a concentration on exploitation, it is moving toward preservation, from sea lion skinners and whalers to sea lion champions and whale watchers. Most important, perhaps most precariously, Patagonia's political leaders are personally collaborating in far-sighted approaches to the enormously difficult challenge of saving nature for those too young to vote.

I sit alone on the brightly colored pebble shores of Punta Tombo that I love so well, on one of the same beaches I visited back in 1964 with Bob, Bill, and Kix and countless times since. One by one, penguins move closer and preen quietly by my side, not because they want to be near me, but because, for a moment, the likes of me are not a threat. I am at peace with them and they with me. For a fleeting exhilarating instant, life is filled with hope. I reflect on how far we have come and on the road ahead. I think back on the extraordinary happenings of Act I and fret over the tragedies of Act II. I boil at the frustrations of Act III, and, finally, I hold my breath in anticipation over what will come as the curtain rises to disclose the next part of the play. How will the stage be set? What parts will be played? Who will take on the leading roles? And who will be in the audience? The nearest penguin sneezes, shakes its head to rid itself of salt, then settles quietly back and closes its eyes. It is dozing. And the southwest Atlantic rolls and hisses a few feet away.

What will best ensure that our children and our children's children can continue to walk in delight within a "nation" of penguins that extends as far as the eye can see; watch in awe as bachelor sea lions raid beachmaster harems; sit spellbound in a small boat next to one of the largest animals Earth has ever known and sense her deep organ-like breathing as though it is their own? How might it be possible for the adventuresome to enjoy the aeronautical perfection of an oceanic albatross in flight over the chilly sea or the incomparable beauty of altiplano flamingos winging between the cold Andean peaks—for our families to watch barranquero families wheeling in color-coordinated perfection before their lofty cliff-edge cities? Might we yet nourish our lives and those of our children by sharing with them the emotional and spiritual experience of a thousand guanacos cantering over the desolate grandeur of the great steppes past flocks of browsing choiques and pairs of faithful maras, beneath condors circling over a puma kill in a vast Tehuelche landscape—and even find green eggs under a bush? Then wild Patagonia will be somewhere.

# Notes

Introduction: *To Patagonia and the Southern Cone*

1. C. Darwin, *The Voyage of H.M.S. Beagle* (1845; repr., Norwalk, Conn.: Heritage Press, 1957), 459.
2. T. Flannery, *The Eternal Frontier* (New York: Atlantic Monthly Press, 2001).
3. E. O. Wilson, *The Future of Life* (New York: Alfred A. Knopf, 2002).

Prelude

1. S. Taber, *Dusk on the Campo: A Journey in Patagonia* (New York: Henry Holt, 1991).
2. T. Flannery, *The Eternal Frontier* (New York: Atlantic Monthly Press, 2001).
3. F. Erize, M. Canevari, P. Canevari, G. Costa, and M. Rumboll, *The National Parks of Argentina and Its Other Natural Areas* (Buenos Aires: Parques Nacionales, 1995).
4. C. Campagna, "Super Seals," *Wildlife Conservation* 4 (1992): 22–27.
5. W. Conway, "Counting Elephant Seals on Punta Norte," *Animal Kingdom* 68, no. 1 (1965): 2–12.
6. L. Brown, *Eco-Economy, Building an Economy for the Earth* (New York: W.W. Norton, 2001).
7. A. Samagalski, *Argentina: A Travel Survival Kit* (Berkeley, Calif.: Lonely Planet, 1989).
8. Erize et al., *National Parks of Argentina*.
9. W. Newmark, "A Land-Bridge Island Perspective on Mammalian Extinction in Western North American Parks," *Nature* 325 (January 29, 1987): 430–432.
10. W. Conway, "Wildlife Conservation Society Patagonian Coast Projects: Examples from a Long-Term Conservation Program," in *AZA Field Conservation Resource Guide* (Silver Spring, Md.: American Zoo and Aquarium Association: 2001), 27–35.
11. G. Harris, *A Guide to the Birds and Mammals of Coastal Patagonia* (Princeton, N.J.: Princeton University Press, 1998).
12. G. Harris, "Animals at Our Door: A Patagonian Family Album," *Animal Kingdom* 92, no. 1 (1989): 12–19.

13. Ibid.
14. Ibid.

Act I: *12,000 Years in Patagonia*

1. P. Szuromi, "Irregular Rise," in "This Week in Science," *Science* 295 (2002): 2321.
2. J. Bird, "Antiquity and Migrations of the Early Inhabitants of Patagonia," *Geographical Review*, no. 28 (1938): 250–275.
3. S. Aramayo and T. Manera de Bianco, "Edad y nuevos hallazgos de icnitas de mamiferos y aves en el yacimiento paleoicnologico de Pehuen-Co (Pleistocene tardio)," *Acociación Paleontológica Argentina de Icnología, Publicación Especial 4, 1 Reunión Argentina de Icnología* (1996): 47–57.
4. Jared Diamond, *Guns, Germs and Steel* (New York: W.W. Norton, 1997), 45.
5. E. O. Wilson, *The Future of Life* (New York: Alfred A. Knopf, 2002).
6. T. Flannery, *The Eternal Frontier* (New York: Atlantic Monthly Press, 2001).
7. Bird, "Antiquity and Migrations"; P. Martin and R. Klein, *Quarternary Extinctions: A Prehistoric Revolution* (Tucson: University of Arizona Press, 1984); Flannery, *The Eternal Frontier*.
8. P. Ward, *Future Evolution* (New York: Times Books, Henry Holt and Company, 2001).
9. D. MacDonald, ed., *The Encyclopedia of Mammals* (New York: Facts on File, 1984).
10. D. Grasso, *Argentina Indigena y Prehistoria Americana*, 3rd ed. (Buenos Aires: Tipografica Editora Argentina, 1981).
11. C. Darwin, *The Voyage of H.M.S. Beagle* (1845; repr., Norwalk, Conn.: Heritage Press, 1957), 210.
12. Ramón Lista, *The Tehuelche Indians: A Disappearing Race*, trans. E. Costa and G. Wise (1894; repr., Argentina: Editorial Confluencia, 1999).
13. G. Williams, *The Desert and the Dream: A Study of Welsh Colonization in Chubut, 1865–1915* (Cardiff: University of Wales Press, 1975).
14. G. C. Musters, *At Home with the Patagonians* (1897; repr., New York: Greenwood Press, 1969), xvi.
15. R. C. Murphy, *Oceanic Birds of South America*, vols. 1 and 2 (New York: Macmillan Company and American Museum of Natural History, 1936), 440.
16. Musters, *At Home with the Patagonians*; F. Reisenberg, *Cape Horn* (New York: Dodd, Mead and Company, 1939); J. Scobie, *Argentina: A City and a Nation* (New York: Oxford University Press, 1964).
17. Musters, *At Home with the Patagonians.*
18. K. Lange, "Jamestown," National Geographic 201, no. 6 (2002): 74-81.

19. Scobie, *Argentina*.
20. Musters, *At Home with the Patagonians*.
21. R. Southwood, *The Story of Life* (New York: Oxford University Press, 2003).
22. C. Darwin, *The Voyage of H.M.S. Beagle* (1845; repr., Norwalk, Conn.: Heritage Press, 1957).

Act II: *Hunting's High Tide*

1. M. Gavirati, "Un Negocio Liviano? La Importancia del Comercio de Plumas de Avestruz para la Colonia Galesa, la Patagonia y Argentina" (unpublished manuscript, 11 pp., CENPAT-CONICET, B. Brown s/n., 9120-Puerto Madryn, Argentina).
2. J. Cajal, "The Lesser Rhea in the Argentine Puna Region: Present Situation," *Biological Conservation* 45 (1988): 81–91.
3. J. C. Godoy, *Fauna Silvestre: Evaluación de los Recursos Naturales de la Argentina*, book 8, vol. 1 (Buenos Aires: Consejo Federal de Inversiones, 1963).
4. K. Raedeke, *Population Dynamics and Socioecology of the Guanaco (*Lama guanicoe*) of Magellanes, Chile* (dissertation, University of Washington, 1979).
5. W. Franklin and M. Fritz, "Sustained Harvesting of the Patagonia Guanaco: Is It Possible or Too Late?" in *Neotropical Wildlife Use and Conservation*, eds. J. Robinson and K. Redford (Chicago: University of Chicago Press, 1991), 317–336.
6. A. Soriano, "Deserts and Semi-Deserts of Patagonia," in *Ecosystems of the World: Temperate Deserts and Semi-Deserts*, ed. Neil West (New York: Elsevier Scientific Publishing, 1983), 423–460.
7. C. Campagna and H. Cappozzo, "Una Historia Trágica," *Vida Silvestre*, 19 (1986): 14–21.
8. G. G. Simpson, *Discoverers of the Lost World* (New Haven, Conn.: Yale University Press, 1984), 92.
9. Ibid.
10. J. Browne, *Charles Darwin, Voyaging* (Princeton, N.J.: Princeton University Press, 1995), 220.
11. F. Erize, M. Canevari, P. Canevari, G. Costa, and M. Rumboll, *The National Parks of Argentina and Its Other Natural Areas* (Buenos Aires: Parques Nacionales, 1995).

Act III: *The Road to Conservation*

1. F. Erize, M. Canevari, P. Canevari, G. Costa, and M. Rumboll, *The National Parks of Argentina and Its Other Natural Areas* (Buenos Aires: Parques Nacionales, 1995).

*Steppe and Altiplano*

2. G. Harris, "Animals at Our Door: A Patagonian Family Album," *Animal Kingdom* 92, no. 1 (1989): 12–19.
3. E. Hoffman and W. Franklin, "The Many Ways Guanacos Talk," *International Wildlife* 25, no. 4 (1993): 4–11.
4. H. Prichard, *Through the Heart of Patagonia* (New York: D. Appleton and Company, 1902), 83–84.
5. G. C. Musters, *At Home with the Patagonians* (1897; repr., New York: Greenwood Press, 1969), 72–73.
6. C. Darwin, *The Voyage of H.M.S. Beagle* (1845; repr., Norwalk, Conn.: Heritage Press, 1957).
7. Musters, *At Home with the Patagonians*.
8. Prichard, *Through the Heart of Patagonia*, 169.
9. K. Raedeke, *Population Dynamics and Socioecology of the Guanaco (*Lama guanicoe*) of Magellanes, Chile* (dissertation, University of Washington, 1979).
10. R. Casamiquela, O. Mondelo, E. Perea, and M. Beros, *Del Mito a la Realidad* (Viedma, Argentina: Fundación Ameghino, 1991).
11. Harris, "Animals at Our Door."
12. J. Cajal and J. Amaya, *Estado Actual de las Investigaciónes Sobre Camélidos en la Republica Argentina* (Buenos Aires: Dirección Nacional de Fauna Silvestre, Secretaria de Ciencia y Tecnica, Ministerio de Educación y Justicia, 1985).
13. A. Soriano and J. Paruelo, "El Pastoreo Ovino," *Ciencia Hoy* 2, no. 7 (1990): 44–53.
14. G. Grimm, "Poblamiento de los Campos Patagónicos," *El Chubut* 20 (July 7, 1994).
15. D. Aagesen, "Crisis and Conservation at the End of the World: Sheep Ranching in Argentine Patagonia," *Environmental Conservation* 27, no. 2 (2000): 208–215.
16. C. Campagna, R. Baldi, and J. Otero, "Counting Sheep, Er, Guanacos in Patagonia," *Wildlife Conservation* 101, no. 1 (1998): 28–35.
17. Ibid.
18. Aagesen, "Crisis and Conservation."
19. Ibid.
20. L. Rohter, "In Patagonia, Sheep Ranches Get Another Chance," *New York Times*, July 23, 2003.
21. M. Antonini and S. Vinella, "Fine Fibre Production from Argentine Camelids—A Development Perspective," *European Fine Fibre Network*, Occasional Publication no. 6 (Rome, Italy: ENEA Innovation Dept., Biotech. and Agricultural Sector, Animal Production Division, 1997).

22. R. Baldi, A. Pelliza-Sbriller, D. Elston, and S. D. Albon, "High Potential for Competition between Guanacos and Sheep in Patagonia," *Journal of Wildlife Management* 68, no. 4 (2004): 924–938.
23. C. Darwin, *The Voyage of H.M.S. Beagle* (1845; repr., Norwalk, Conn.: Heritage Press, 1957), 152.
24. W. Karesh, M. Uhart, D. Dierenfeld, W. E. Braselton, A. Torres, C. House, H. Puche, and R. Cook, "Health Evaluation of Free-Ranging Guanaco (*Lama guanicoe*)," *Journal of Zoo and Wildlife Medicine* 29, no. 2 (1998): 134–141.
25. K. Lafferty and L. Gerber, "Good Medicine for Conservation Biology: The Intersection of Epidemiology and Conservation Theory," *Conservation Biology* 16, no. 3 (2002): 593–604.
26. Darwin, *The Voyage of H.M.S. Beagle*, 152.
27. D. Bruning, "Social Structure and Reproductive Behavior in the Greater Rhea," in D. Lancaster, ed., *The Living Bird*, 13th annual (1974), 251–294.
28. G. Fernández and J. Reboreda, "Male Parental Care in Greater Rheas (*Rhea americana*) in Argentina," *The Auk* 120, no. 2 (2003): 418–428.
29. P. L. Sclater, *Argentine Ornithology*, vol. 2 (London: R.H. Porter, 1889).
30. G. Dubost and H. Genest, "Le Comportement Social d'une Colonie de Maras (*Dolichotis patagonum*)," *Zeitschrift für Tierpsychologie* 35, no. 3 (1974): 225–302.
31. S. Taber, *Dusk on the Campo: A Journey in Patagonia* (New York: Henry Holt, 1991).
32. D. MacDonald, ed., *The Encyclopedia of Mammals* (New York: Facts on File, 1984).
33. Dubost and Genest, "Le Comportement Social d'une Colonie de Maras."
34. N. Collar, "Priorities for Parrot Conservation in the New World," *Cotinga* 5 (1996): 26–31.
35. Irene Pepperberg, *The Alex Studies* (Cambridge, Mass.: Harvard University Press, 1999).
36. Darwin, *The Voyage of H.M.S. Beagle*, 57.
37. A. Glade, "Proyecto de Conservación y Manejo del Loro Tricahue," Informe Temporada 1984–1985, Doc. de Trabajo no. 55 (Chile: CONAF, 1985).
38. J. Masello and P. Quillfeldt, "Chick Growth and Breeding Success of the Burrowing Parrot," *The Condor* 104 (2002): 574–586; J. Masello, A. Sramkova, P. Quillfeldt, J. Epplen, and T. Lubjuhn, "Genetic Monogamy in Burrowing Parrots (*Cyanoliseus patagonus*)?" *Journal of Avian Biology* 33 (2000): 99–103.

39. Sclater, *Argentine Ornithology*, vol. 2.
40. Masello and Quillfeldt, "Chick Growth and Breeding Success of the Burrowing Parrot."
41. P. Walker, "*Geochelone chilensis*, Chaco Tortoise," in *The Conservation Biology of Tortoises*, Occasional Papers of the IUCN Species Survival Commission (SSC) no. 5, eds. I. Swingland and M. Klemens (Gland, Switzerland: IUCN, 1990), 20–21; T. Waller, "Los Criaderos Como Pantallas para el Comercio Internaciónal de Tortugas en Argentina," *Noticites '87* (Ottawa, Canada) 1, no. 1 (1987): 3–4.
42. A. Salzberg, "Chelonian Conservation News," *Chelonian Conservation and Biology* 3, no. 1 (1998): 147–150.
43. T. Teleck, "United States Import and Export of Live Turtles and Tortoises," *Turtle and Tortoise Newsletter* 4 (2001): 8–13.
44. Darwin, *The Voyage of H.M.S. Beagle.*
45. T. Flannery, *The Eternal Frontier* (New York: Atlantic Monthly Press, 2001).
46. C. Vinci, "El Porque del Area Natural Protegida (Reserva Natural Integral Meseta de Somuncurá)," in *La Meseta Patagónica del Somuncurá*, coordinador R. Freddy, 385–395 (Gobierno del Chubut, Gobierno de Rio Negro, 1993).
47. M. Andrews, *The Flight of the Condor* (Boston: Little, Brown, 1982).
48. C. Wilmers, R. Crabtree, D. Smith, K. Murphy, and W. Getz, "Trophic Facilitation by Introduced Top Predators: Grey Wolf Subsidies to Scavengers in Yellowstone National Park," *Journal of Animal Ecology* 72 (2003): 909–916.
49. F. Hiraldo, J. Donázar, O. Ceballos, A. Travaini, J. Bustamante, and M. Funes, "Breeding Biology of the Grey Eagle–Buzzard Population in Patagonia," *Wilson Bulletin* 107, no. 4 (1995): 675–685.
50. A. Novaro, M. Funes, and S. Walker, "Ecological Extinction of Native Prey of a Carnivore Assemblage in Argentine Patagonia," *Biological Conservation* 92 (2000): 25–33.
51. A. Novaro, "Sustainability of Harvest of Culpeo Foxes in Patagonia," *Oryx* 29, no. 1 (1995): 18–22.
52. "Briefly," *Oryx* 36, no. 4 (2002): 313.
53. D. Pimental, L. Lach, R. Zuniga, and D. Morrison, "Environmental and Economic Costs of Nonindigenous Species in the United States," *BioScience* 50, no. 1 (2000): 53–65.
54. J. Daciuk, "La Fauna del Parque Nacional Laguna Blanca," *Anales de Parques Nacionales*, tomo 11, estrega 2a (1968): 225–304.
55. J. Terborgh, J. Estes, P. Paquet, K. Ralls, D. Boyd-Heger, B. Miller, and R. Noss, "The Role of Top Carnivores in Regulating Terrestrial

Ecosystems," in *Continental Conservation: Scientific Foundations of Regional Reserve Networks*, eds. M. Soulé and J. Terborgh (Washington, D.C.: Island Press, 1999); J. Laundré and T. Clark, "Managing Puma Hunting in the Western United States through a Metapopulation Approach," *Animal Conservation* 6 (2003): 159–170.

56. A. Novaro and S. Walker, "Human-Induced Changes in the Effect of Top Carnivores on Biodiversity in the Patagonian Steppe," in *Large Carnivores and the Conservation of Biodiversity*, eds. J. C. Ray, K. H. Redford, R. S. Steneck, and J. Berger (Washington, D.C.: Island Press, 2005).

57. W. Conway, "To the High Andes for the Rarest Flamingo," *Animal Kingdom* 63, no. 2 (1960): 34–50; W. Conway, *To the High Andes for the Rarest Flamingo*, 22-min. sound & color film photographed by B. Littlehales, directed by W. Conway, edited by E. Anton, New York Zoological Society, 1960; W. Conway, "In Quest of the Rarest Flamingo," *National Geographic Magazine* 120, no. 1 (1961): 91–105.

58. R. Allen, *The Flamingos: Their Life History and Survival*, Research Report no. 5, (New York, National Audubon Society, 1956); A. W. Johnson, *Birds of Chile and Adjacent Regions of Argentina, Bolivia and Peru*, vols. 1 and 2 (Buenos Aires: Platt Establecimientos Gráficos, 1965).

59. O. Rocha, *Contribución Preliminar a la Conservación y el Conocimiento de la Ecología de Flamencos en la Reserva Nacional de Fauna Andina "Eduardo Avaroa"* (La Paz, Bolivia: Departmento Potosí and Museo Nacional de Historia Natural, 1994).

60. H. Berry, "South West Africa," in *Flamingos: Populations, Ecology and the Conservation*, eds. J. Kear and N. Duplaix-Hall (Hertfordshire, England: The Wildfowl Trust and T. & A.D. Poyser, 1975), 53–64.

61. M. P. Kahl, "Distribution and Numbers: A Summary," in *Flamingos: Populations, Ecology and the Conservation*, eds. J. Kear and N. Duplaix-Hall (Hertfordshire, England: The Wildfowl Trust and T. & A.D. Poyser, 1975), 93–102.

62. D. Reed, J. O'Grady, B. Brook, J. Ballou, and R. Frankham, "Estimates of Minimum Viable Population Sizes for Vertebrates and Factors Influencing Those Estimates," *Biological Conservation* 113 (2003): 23–34.

63. C. Elton, *The Ecology of Animals* (London: Methuen & Company, 1933), 27–30.

64. C. Loucks, Z. Lü, E. Dinerstein, H. Wang, D. Olsen, C. Zhu, and D. Wang, "Giant Pandas in a Changing Landscape," *Science* 294 (2001): 1465.

65. N. Bonino, G. Bonssuvito, and A. Sbriller, "Composición Botánica de la Dieta de Herbivoros Silvestres y Domésticos en el Áreas de Pilcaniyeu

(Rio Negro) 2: Calculo de los Índice de Diversidad Trófica y Similitud," *Communicación Técnica*, no. 10 (San Carlos de Bariloche, Argentina: INTA, 1986).

66. R. Raskin, B. Alexander, J. van den Noort, and R. Carter, *Prosperity in the 21st Century West: The Role of Protected Public Lands* (Tucson: Ariz.: Sonoran Institute, July 2004).

*Coastal Chronicles*

67. C. Campagna, *The Breeding Behavior of the Southern Sea Lion* (dissertation, University of California, Santa Cruz, 1987).
68. K. Kendrick, A. P. Da Costa, A. E. Leigh, M. R. Hinton, and J. W. Peirce, "Sheep Don't Forget a Face," *Nature* 414 (November 8, 2001): 165.
69. M. Riedman, *The Pinnipeds, Seals, Sea Lions, and Walruses* (Berkeley: University of California Press, 1990).
70. R. Bastida and D. Rodriquez, *Mamiferos Marinos de Patagonia y Antartida* (Buenos Aires: Vazquez Mazzini Editores, 2003).
71. W. B. Karesh, *Appointment at the Ends of the World* (New York: Warner Books, 1999).
72. D. Thompson and C. Duck, *Southern Sea Lions (*Otaria flavescens*) in the Falkland Islands* (Cambridge, England: Natural Environment Research Council, Sea Mammal Research Unit, 1995).
73. G. Dehnhardt, B. Mauck, W. Hanke, and H. Bleckman, "Hydrodynamic Trail Following in Harbor Seals (*Phoca vitulina*)," *Science* 293 (2001): 102–104.
74. Riedman, *The Pinnipeds, Seals, Sea Lions, and Walruses*.
75. C. Campagna, "Movements and Location at Sea of South American Sea Lions (*Otaria flavescens*)," *Journal of Zoology* 257 (2001): 205–220.
76. E. Crespo, S. Pedraza, S. Dans, M. Alonso, A. L. Reyes, N. Garcia, M. Coscarella, and A. Schiavini, "Direct and Indirect Effects of the High-seas Fisheries on the Marine Mammal Populations in the Northern and Central Patagonian Coast," *Journal of Northwest Atlantic Fishery Science* 22 (1997): 189–207.
77. T. Jensen, M. van de Bildt, H. Dietz, T. Andersen, A. Hammer, T. Kuiken, and A. Osterhaus, "Another Phocine Distemper Outbreak in Europe," *Science* 297 (2002): 209.
78. Crespo et al., "Direct and Indirect Effects of the Highseas Fisheries."
79. P. Marquet, "Of Predators, Prey, and Power Laws," *Science* 295 (2002): 2229–2230.
80. C. Carbone and J. Gittleman, "A Common Rule for the Scaling of Carnivore Density," *Science* 295 (2002): 2273–2276.
81. Crespo et al., "Direct and Indirect Effects of the Highseas Fisheries."

82. C. Campagna, "Super Seals," *Wildlife Conservation* 4 (1992): 22–27.
83. B. Le Boeuf, "Incredible Diving Machines," *Natural History* 2 (1989): 34–41.
84. C. Campagna, *Sobre la Foca Elefante: Historias Naturales de la Patagonia* (Buenos Aires: Fondo de Cultura Económica de Argentina, 2002).
85. F. Ashcroft, *Life at the Extremes* (Berkeley: University of California Press, 2000).
86. Le Boeuf, "Incredible Diving Machines."
87. C. Harding, "Going to Extremes," *National Wildlife* (Aug/Sep 1993): 39–44.
88. Le Boeuf, "Incredible Diving Machines."
89. C. Campagna, "Diving Behavior and Foraging Location of Female Southern Elephant Seals from Patagonia," *Journal of Zoology* (London) 236 (1995): 55–71.
90. Campagna, "Movements and Location at Sea of South American Sea Lions."
91. T. Williams, R. Davis, L. Fuiman, J. Francis, B. Le Boeuf, M. Horning, J. Calambokidis, and D. Croll, "Sink or Swim: Strategies for Cost-Efficient Diving by Marine Mammals," *Science* 288 (2000): 133–136.
92. Le Boeuf, "Incredible Diving Machines."
93. D. Wilcove, *The Condor's Shadow* (New York: W.H. Freeman, 1999).
94. A. Hoelzel, J. Halley, S. O'Brien, C. Campagna, T. Arnbom, B. Le Boeuf, K. Ralls, and G. Dover, "Elephant Seal Genetic Variation and the Use of Simulation Models to Investigate Historical Population Bottlenecks," *Journal of Heredity* 84, no. 6 (2003): 443–449.
95. Campagna, *Sobre la Foca Elefante*.
96. Riedman, *The Pinnipeds, Seals, Sea Lions, and Walruses*.
97. J. Reader and H. Croze, *Pyramids of Life: Illuminations of Nature's Fearful Symmetry* (Philadelphia: J.B. Lippincott, 1977).
98. S. Pimm, *The World According to Pimm* (New York: McGraw-Hill, 2001).
99. A. W. Johnson, *Birds of Chile and Adjacent Regions of Argentina, Bolivia and Peru*, vols. 1 and 2 (Buenos Aires: Platt Establecimientos Gráficos, 1965).
100. Ibid., 126.
101. H. Durnford, "Notes on the Birds of Central Patagonia," *Ibis* 4, no. 2 (1878): 389–406.
102. D. Duffy and W. Siegfried, "Historical Variations in Food Consumption by Breeding Seabirds of the Humboldt and Benguela Upwelling Regions," in *Seabirds: Feeding Biology and Role in Marine Ecosystems*, ed. J. Croxall (Cambridge: Cambridge University Press, 1987), 327–346.
103. R. C. Murphy, *Oceanic Birds of South America*, vols. 1 and 2 (New York: Macmillan Company and American Museum of Natural History, 1936).

104. P. Majluf, "The Poop on Penguins," *Wildlife Conservation* 105, no. 4 (2002): 30–33.
105. G. Winegrad, "Memo to American Bird Conservancy Members, May 5, 2003" (ABC, 1834 Jefferson Place NW, Washington, DC 20036); G. Winegrad, "Opposition Grows to Cormorant Slaughter," Bird Calls 7, no. 2 (2003): 3.
106. J. C. Godoy, *Fauna Silvestre: Evaluación de los Recursos Naturales de la Argentina*, tomo 8, vol. 1 (Buenos Aires: Consejo Federal de Inversiones, 1963).
107. D. Grémillet, "Foraging Energetics of Arctic Cormorants and the Evolution of Diving Birds," *Ecological Letter* 4 (2001): 180–184.
108. A. Sapoznikow and F. Quintana, "Foraging Behavior and Feeding Locations of Imperial Cormorants and Rock Shags Breeding Sympatrically in Patagonia, Argentina," *Waterbirds* 26, no. 2 (2003): 184–191.
109. G. Simpson, *Penguins, Past and Present, Here and There* (New Haven, Conn.: Yale University Press, 1976).
110. Durnford, "Notes on the Birds of Central Patagonia."
111. D. Renison, D. Boersma, and M. Martella, "Winning and Losing: Causes for Variability in Outcome of Fights in Male Magellanic Penguins (*Spheniscus magellanicus*)," *Behavioral Ecology* 13, no. 4 (2002): 462–466.
112. P. Yorio, E. Frere, P. Gandini, and G. Harris, *Atlas de la Distribución Reproductiva de Aves Marinas en el Litoral Patagónico Argentino* (Puerto Madryn, Chubut, Argentina: Fundación Patagonia Natural; New York: Wildlife Conservation Society, 1998).
113. M. Riedman, "The Evolution of Alloparental Care and Adoption in Mammals and Birds," *Quarterly Review of Biology* 57 (1982): 405–435.
114. R. Wilson and M.-P. Wilson, "Foraging Ecology of Breeding Penguins," in *Penguin Biology*, eds. L. Davis and J. Darby (New York: Academic Press, 1990), 181–206.
115. B. Walker and P. D. Boersma, "Diving Behavior of Magellanic Penguins (*Spheniscus magellanicus*) at Punta Tombo, Argentina," *Canadian Journal of Zoology* 81 (2003): 1471–1483.
116. T. Burnham and J. Phelan, *Mean Genes* (Cambridge, Mass.: Perseus Publishing, 2000).
117. T. Coulsen, J. Lindström, and P. Cotgreave, "Seeking New Recruits," *Science* 295 (2003): 2023–2024.
118. C. Safina, *Eye of the Albatross* (New York: Henry Holt, 2002).
119. C. Rolland, E. Danchin, and M. de Fraipont, "The Evolution of Coloniality in Birds in Relation to Food, Habitat, Predation, and Life History Traits: A Comparative Analysis," *American Naturalist* 151 (1998):

414–529; A. Hernandez-Matias, L. Jover, and X. Ruis, "Predation of Common Tern Eggs in Relation to Sub-Colony Size, Nest Aggregation, and Breeding Synchrony," *Waterbirds* 26, no. 3 (2003): 280–289.

120. C. Campagna, C. Bisioli, F. Quintana, and A. Vila, "Group Breeding Sea Lions: Pups Survive Better in Colonies," *Animal Behavior* 43 (1992): 541–548.
121. R. Moss, S. Wanless, and M. Harris, "How Small Northern Gannet Colonies Grow Faster Than Big Ones," *Waterbirds* 25, no. 4 (2002): 442–448.
122. P. Yorio, F. Quintana, C. Campagna, and G. Harris, "Ecology and Conservation of Seabirds and Marine Mammals at Punta León: Final Report," 1993 (New York: Wildlife Conservation International, unpublished).
123. P. Yorio and F. Quintana, "Predation by Kelp Gulls *Larus dominicanus* at a Mixed Species Colony of Royal Terns *Sterna maxima* and Cayenne Terns *Sterna eurygnatha* in Patagonia," *Ibis* 139 (1997): 536–541.
124. Yorio et al., "Ecology and Conservation of Seabirds and Marine Mammals at Punta León."
125. Safina, *Eye of the Albatross.*
126. Fundación Patagonia Natural, news release, 2004.
127. A. Lichter, ed., *Tracks in the Sand, Shadows in the Sea* (Buenos Aires: Ediciónes Terra Nova, 1992).
128. W. Conway, "On the Shores of a Cold Sea River," in *Tracks in the Sand, Shadows in the Sea*, ed. Alfredo Lichter (Buenos Aires: Ediciónes Terra Nova, 1992), 164–172.
129. J. Croxall and A. Wood, "The Importance of the Patagonian Shelf for Top Predator Species Breeding at South Georgia," *Aquatic Conservation: Marine and Freshwater Ecosystems* 12 (2002): 101–118.

*Sea and Sky*

130. G. Caille, R. Gonzalez, A. Gostanyi, and N. Ciocco, "Especies Capturadas por las Flotas de Pesca Costera en Patagonia. Programa de Biologos Observadores a Bordo, 1993–1996," *Informes Técnicos del Plan de Manejo Integrado de Zona Costera Patagónica* (Fundación Patagonia Natural, Puerto Madryn, Argentina), no. 27 (2002): 1–21.
131. S. Norris, M. Hall, E. Melvin, and J. Parrish, "Thinking Like an Ocean," *Conservation in Practice* 3, no. 4 (2002): 10–19.
132. P. Ward, *Future Evolution* (New York: Times Books, Henry Holt and Company, 2001).
133. Norris et al., "Thinking Like an Ocean."
134. G. Robertson, "Effect of Line Sink Rate on Albatross Mortality in the

Patagonian Toothfish Longline Fishery, in *Seabird Bycatch: Trends, Roadblocks and Solutions*, eds. E. Melvin and J. Parrish (University of Alaska Sea Grant, 2001), 43–60.

135. Food and Agriculture Organization of the United Nations, *Review of the State of World Fishery Resources*, FAO Fisheries Circular no. 920 FIRM/C920, Rome, Italy.
136. D. Laffoley, "Techniques for Managing Marine Protected Areas: Zoning," in *Marine Protected Areas: Principles and Techniques for Management*, ed. S. Gubbay (London: Chapman and Hall, 1995), 102–118.
137. C. Roberts, B. Halpern, S. Palumbi, and R. Warner, "Designing Marine Reserve Networks," *Conservation Biology in Practice* 2, no. 3 (2001): 10–17; F. Gell and C. Roberts, "Benefits beyond Boundaries: The Fishery Effects of Marine Reserves," *TREE* 18, no. 9 (2003): 448–455.
138. B. Stempeck, *Greenwire*, www.greenwire.com, October 28, 2002.
139. D. Pauly, "The Crisis in Fisheries and Marine Biodiversity," contribution to *Sustaining Seascapes: The Science and Policy of Marine Resources Management*, American Museum of Natural History, New York, March 7–8, 2002.
140. J. Robinson and E. Bennett, eds., *Hunting for Sustainability in Tropical Forests* (New York: Columbia University Press, 2000).
141. C. Waluda, P. Trathan, C. Elvidge, V. Hobson, and P. Rodhouse, "Throwing Light on Straddling Stocks of *Illex argentinus*: Assessing Fishing Intensity with Satellite Imagery," *Canadian Journal of Fisheries and Aquatic Science* 59 (2002): 592–596.
142. Food and Agriculture Organization of the United Nations, *Review of the State of World Fishery Resources*.
143. R. Myers and B. Worm, "Rapid Worldwide Depletion of Predatory Fish Communities," *Nature* 423 (May 15, 2003): 280–283.
144. Pimm, *The World According to Pimm*.
145. J. Jackson et al., "Historical Overfishing and the Recent Collapse of Coastal Ecosystems," *Science* 293 (2001): 629–638.
146. Ibid.
147. Ibid.
148. "Letters to the Editor," *Time*, 1988.
149. B. Estabrook, "The Wild and the Farmed," *Gourmet* (September 2002): 90–94.
150. Ibid.
151. Fen Montaigne, "Atlantic Salmon," *National Geographic* 204, no. 1 (July 2003): 100–123.
152. *Economist* (August 9, 2003): 19–21.
153. G. Gajardo and L. Laikre, "Chilean Aquaculture Boom Is Based on

Exotic Salmon Resources: A Conservation Paradox," *Conservation Biology* 17, no. 4 (2003): 1173–1174.
154. Food and Agricultural Organization of the United Nations, *Food Balance Sheets* (Rome, Italy: Author, 1991).
155. Pimm, *The World According to Pimm.*
156. Bastida and Rodriquez, *Mamiferos Marinos de Patagonia y Antartida.*
157. R. Ellis, *Dolphins and Porpoises* (New York: Alfred A. Knopf, 1982).
158. C. Campagna and J. C. Lopez, "Sea Pandas," *Wildlife Conservation* 4 (1994): 44–51.
159. Ellis, *Dolphins and Porpoises.*
160. J. Rossi, *The Wild Shores of Patagonia* (New York: Harry N. Abrams, 2000).
161. Ellis, *Dolphins and Porpoises.*
162. J. C. Lopez, *Orcas, Entre el Mito y la Realidad* (Buenos Aires: Rumbo Sur, Editorial Sudamerican, 2000).
163. Pimm, *The World According to Pimm.*
164. Bastida and Rodriquez, *Mamiferos Marinos de Patagonia y Antartida.*
165. R. Payne, *Among Whales* (New York: Scribner, 1995).
166. G. Harris and C. Garcia, *The Right Whales of Peninsula Valdés* (Puerto Madryn, Chubut, Argentina: Impresora, 1986).
167. Payne, *Among Whales.*
168. C. Campagna and A. Lichter, *Las Ballenas de la Patagonia* (Buenos Aires: emecé editores, 1996).
169. D. Attenborough, "The Blue Planet: Seas of Life," BBC and Discovery Channel, 2002 (now on DVD).
170. C. Clark, "The Economics of Overexploitation," *Science* 181 (1973): 630–633.
171. A. Dobson, *Conservation and Biodiversity* (New York: Scientific American Library, 1996).
172. E. O. Wilson, *The Future of Life* (New York: Alfred A. Knopf, 2002).
173. Modified after Dobson, *Conservation and Biodiversity.*
174. V. F. Boyson, *The Falkland Islands* (London: Oxford University Press, 1924).
175. R. Woods, *Guide to the Birds of the Falkland Islands* (Shropshire, England: Anthony Nelson, 1988); I. Strange, *A Field Guide to the Wildlife of the Falkland Islands and South Georgia* (London: HarperCollins, 1992); D. Summers, *A Visitor's Guide to the Falkland Islands* (Finchley, London: Falklands Conservation, 2001).
176. Woods, *Guide to the Birds of the Falkland Islands.*
177. Murphy, *Oceanic Birds of South America.*
178. Safina, *Eye of the Albatross.*

179. K. Putz, A. Clausen, N. Huin, and J. Croxall, "Re-evaluation of Historical Rockhopper Penguin Population Data in the Falkland Islands," *Waterbirds* 26, no. 2 (2003): 169–175.
180. Strange, *A Field Guide to the Wildlife of the Falkland Islands.*
181. Safina, *Eye of the Albatross.*
182. B. Mearns and R. Mearns, *The Bird Collectors* (San Diego, Calif.: Academic Press, 1998).
183. Safina, *Eye of the Albatross.*
184. Ibid.
185. G. Shire, "Baiting the Hook to Suit the Fish," *Bird Conservation* 17 (2001): 8–9.
186. N. Huin, "Dramatic Declines in Steeple Jason Seabird Populations," *Falklands Conservation Newsletter* (December 2003): 1.

The Scene Ahead

1. *2004 IUCN Red List of Threatened Species.* Constantly updated, this list is best viewed on the Internet at www.redlist.org.
2. Y. Martel, *Life of Pi* (New York: Harcourt, 2001).
3. E. O. Wilson, *The Future of Life* (New York: Alfred A. Knopf, 2002).
4. IUCN, *Parks for Life. Report of the Fourth World Congress on National Parks and Protected Areas* (Gland, Switzerland: Author, 1993).
5. E. Larson, *Evolution's Workshop* (New York: Basic Books, 2001).
6. M. Soulé, ed., *Conservation Biology: The Science of Scarcity and Diversity* (Sunderland, Mass.: Sinauer Associates, 1986).
7. F. Erize, M. Canevari, P. Canevari, G. Costa, and M. Rumboll, *Los Parques Nacionales de la Argentina* (Buenos Aires: Parques Nacionales, 1981).
8. Wilson, *The Future of Life.*
9. C. Meine, *Correction Lines: Essays on Land, Leopold, and Conservation* (Washington, D.C.: Island Press, 2004).

# Suggested Reading

Several English-language books are available for those traveling to the Southern Cone, but some of the best references are in Spanish; and some of those in either language are hard to find. Those most easily acquired are marked with an asterisk in the list below.

The World Wide Web now contains a rich trove of Southern Cone information and pictures, including listings of conservation organizations, along with the customary travel promotions. Fundación Patagonia Natural (FPN), Fundación Vida Silvestre Argentina, Aves Argentina, and others have useful Web sites and, in some instances, discussions of current conservation concerns, as does the Web site of the Wildlife Conservation Society in New York. The FPN Web site includes an excellent handbook based on the UNDP/WCS-sponsored Patagonian Coastal Zone Management Plan. Under "Sea and Sky, Patagonia," an up-to-date discussion of that exceptionally important project is provided.

*Bastida, R., and D. Rodriquez. 2003. *Mamiferos Marinos de Patagonia y Antartida*. Vazquez Mazzini Editores, Buenos Aires. Well illustrated and the most up-to-date guide on marine mammals in the Southern Cone.

Campagna, C. 2002. *Sobre la Foca Elefante: Historias Naturales de la Patagonia*. Fondo de Cultura Económica de Argentina, Buenos Aires. A deeply thoughtful and personal discussion of elephant seals from one of the top researchers.

*Darwin, C. (1845 version.) *The Voyage of H.M.S. Beagle*. Heritage Press, Norwalk, Connecticut. The classic—and an excellent travel companion; paperback available.

*De La Peña, M., and M. Rumboll. 1998. *Birds of Southern South America and Antarctica*. Princeton University Press, Princeton, New Jersey. A pocket-sized checklist for identifications, exceptionally well illustrated.

Erize, F., Canevari, M., Canevari, P., Costa, G., and M. Rumboll. 1981. *Los Parques Nacionales de la Argentina*. Parques Nacionales, Buenos Aires. This large-format book and the English version below contain an exceptional amount of information and excellent photographs.

Erize, F., Canevari, M., Canevari, P., Costa, G., and M. Rumboll. 1995. *The*

*National Parks of Argentina and Its Other Natural Areas*. Parques Nacionales, Buenos Aires.

*Harris, G. 1998. *A Guide to the Birds and Mammals of Coastal Patagonia*. Princeton University Press, Princeton, New Jersey. The best-illustrated and most informative guide to the wildlife of the Patagonian Coast.

Murphy, R. 1936. *Oceanic Birds of South America*, vols. 1 and 2. Macmillan Company and American Museum of Natural History, New York. A two-volume opus filled with wonderful material, for serious reading at home. Hard to find.

Musters, G. C. 1897. *At Home with the Patagonians*. Greenwood Press Reprinting, 1969, Greenwood Press, New York. About the Tehuelche Indians; hard to find but absolutely fascinating.

*Parera, A., and F. Erize. 2002. *Los Mamiferos de la Argentina y la Region Austral de Sudamerica, 1st ed. El Ateneo, Buenos Aires.* A large-format, richly illustrated book covering all the known mammals of Argentina and nearby. Excellent.

Rossi, J. 2000. *The Wild Shores of Patagonia.* Harry N. Abrams, New York. A large-format photographic presentation of Peninsula Valdés and Punta Tombo; beautiful photographs and an excellent text.

*Safina, C. 2002. *Eye of the Albatross.* Henry Holt, New York. This sizable, prize-winning book is a beautifully written account of some of the most fascinating creatures alive. Big to carry on a trip but worth it.

*Strange, I. 1992. *A Field Guide to the Wildlife of the Falkland Islands and South Georgia.* HarperCollins, London. Excellent illustrated, pocket-sized guide to almost every living thing in the Falkland/Malvinas.

*Summers, D. 2001. *A Visitor's Guide to the Falkland Islands.* Falklands Conservation, Finchley, London. Handy island-by-island coverage, with maps and descriptions.

Woods, R. 1988. *Guide to the Birds of the Falkland Islands.* Anthony Nelson, Shropshire, England. Excellent detailed bird guide to the islands.

# Index

# Index

# Index

# Index

## Index

# Index